Crosswalk Coach
for the Common Core State Standards

Mathematics

Grade 8

Crosswalk Coach for the Common Core State Standards, Mathematics, Grade 8
303NA
ISBN-13: 978-0-7836-7852-8

Contributing Writer: Colleen O'Donnell Oppenzato
Cover Image: © Ron Hilton/Dreamstime.com

Triumph Learning® 136 Madison Avenue, 7th Floor, New York, NY 10016

Frequently Asked Questions about the Common Core State Standards

What are the Common Core State Standards?

The Common Core State Standards for mathematics and English language arts, grades K–12, are a set of shared goals and expectations for the knowledge and skills that will help students succeed. They allow students to understand what is expected of them and to become progressively more proficient in understanding and using mathematics and English language arts. Teachers will be better equipped to know exactly what they must do to help students learn and to establish individualized benchmarks for them.

Will the Common Core State Standards tell teachers how and what to teach?

No. Because the best understanding of what works in the classroom comes from teachers, these standards will establish *what* students need to learn, but they will not dictate *how* teachers should teach. Instead, schools and teachers will decide how best to help students reach the standards.

What will the Common Core State Standards mean for students?

The standards will provide a clear, consistent understanding of what is expected of student learning across the country. Common standards will not prevent different levels of achievement among students, but they will ensure more consistent exposure to materials and learning experiences through curriculum, instruction, teacher preparation, and other supports for student learning. These standards will help give students the knowledge and skills they need to succeed in college and careers.

Do the Common Core State Standards focus on skills and content knowledge?

Yes. The Common Core Standards recognize that both content and skills are important. They require rigorous content and application of knowledge through higher-order thinking skills. The English language arts standards require certain critical content for all students, including classic myths and stories from around the world, America's founding documents, foundational American literature, and Shakespeare. The remaining crucial decisions about content are left to state and local determination. In addition to content coverage, the Common Core State Standards require that students systematically acquire knowledge of literature and other disciplines through reading, writing, speaking, and listening.

In mathematics, the Common Core State Standards lay a solid foundation in whole numbers, addition, subtraction, multiplication, division, fractions, and decimals. Together, these elements support a student's ability to learn and apply more demanding math concepts and procedures.

The Common Core State Standards require that students develop a depth of understanding and ability to apply English language arts and mathematics to novel situations, as college students and employees regularly do.

Will common assessments be developed?

It will be up to the states: some states plan to come together voluntarily to develop a common assessment system. A state-led consortium on assessment would be grounded in the following principles: allowing for comparison across students, schools, districts, states and nations; creating economies of scale; providing information and supporting more effective teaching and learning; and preparing students for college and careers.

Table of Contents

Common Core State Standards Correlation Chart 6

			Common Core State Standards
Domain 1	**The Number System**	11	
Domain 1: Diagnostic Assessment for Lessons 1–4	12	
Lesson 1	Rational Numbers	14	8.NS.1
Lesson 2	Irrational Numbers	20	8.NS.1, 8.NS.2, 8.EE.2
Lesson 3	Compare and Order Rational and Irrational Numbers	26	8.NS.1, 8.NS.2
Lesson 4	Estimate the Value of Expressions	31	8.NS.2
Domain 1: Cumulative Assessment for Lessons 1–4	36	
Domain 2	**Expressions and Equations**	39	
Domain 2: Diagnostic Assessment for Lessons 5–18	40	
Lesson 5	Exponents	43	8.EE.1
Lesson 6	Square Roots and Cube Roots	49	8.EE.2
Lesson 7	Scientific Notation	54	8.EE.4
Lesson 8	Solve Problems Using Scientific Notation	62	8.EE.3, 8.EE.4
Lesson 9	Linear Equations in One Variable	67	8.EE.7.a, 8.EE.7.b
Lesson 10	Use One-Variable Linear Equations to Solve Problems	75	8.EE.7.a, 8.EE.7.b
Lesson 11	Slope	80	8.EE.6
Lesson 12	Slopes and *y*-intercepts	88	8.EE.5, 8.EE.6
Lesson 13	Proportional Relationships	95	8.EE.5
Lesson 14	Direct Proportions	101	8.EE.5
Lesson 15	Pairs of Linear Equations	107	8.EE.8
Lesson 16	Solve Systems of Equations Graphically	113	8.EE.8.a, 8.EE.8.b
Lesson 17	Solve Systems of Equations Algebraically	120	8.EE.8.a, 8.EE.8.b, 8.EE.8.c
Lesson 18	Use Systems of Equations to Solve Problems . .	128	8.EE.8.c
Domain 2: Cumulative Assessment for Lessons 5–18	133	
Domain 3	**Functions**	137	
Domain 3: Diagnostic Assessment for Lessons 19–23	138	
Lesson 19	Introduction to Functions	141	8.F.1, 8.F.3
Lesson 20	Work with Linear Functions	149	8.F.3, 8.F.4

			Common Core State Standards
Lesson 21	Use Functions to Solve Problems 155		8.F.4
Lesson 22	Use Graphs to Describe Relationships 163		8.F.5
Lesson 23	Compare Relationships Represented in Different Ways 169		8.EE.5, 8.F.2

Domain 3: Cumulative Assessment for Lessons 19–23 176

Domain 4   Geometry 179
Domain 4: Diagnostic Assessment for Lessons 24–32 180

Lesson 24	Congruence Transformations 183		8.G.1.a, 8.G.1.b, 8.G.1.c, 8.G.2, 8.G.3
Lesson 25	Dilations 190		8.G.3, 8.G.4
Lesson 26	Similar Triangles 196		8.G.4, 8.G.5
Lesson 27	Interior and Exterior Angles of Triangles 205		8.G.5
Lesson 28	Parallel Lines and Transversals 211		8.G.5
Lesson 29	The Pythagorean Theorem 218		8.G.6, 8.G.7
Lesson 30	Distance 224		8.G.8
Lesson 31	Apply the Pythagorean Theorem 230		8.G.7, 8.G.8
Lesson 32	Volume 236		8.G.9

Domain 4: Cumulative Assessment for Lessons 24–32 242

Domain 5   Statistics and Probability 245
Domain 5: Diagnostic Assessment for Lessons 33–36 246

Lesson 33	Scatter Plots 249		8.SP.1
Lesson 34	Trend Lines 257		8.SP.1, 8.SP.2
Lesson 35	Interpret Linear Models 263		8.SP.3
Lesson 36	Patterns in Data 269		8.SP.4

Domain 5: Cumulative Assessment for Lessons 33–36 277

Glossary 280
Summative Assessment: Domains 1–5 285
Math Tools 305

Common Core State Standards Correlation Chart

Common Core State Standard	Grade 8	*Coach Lesson(s)*
Domain: The Number System		
Know that there are numbers that are not rational, and approximate them by rational numbers.		
8.NS.1	Understand informally that every number has a decimal expansion; the rational numbers are those with decimal expansions that terminate in 0s or eventually repeat. Know that other numbers are called irrational.	1–3
8.NS.2	Use rational approximations of irrational numbers to compare the size of irrational numbers, locate them approximately on a number line diagram, and estimate the value of expressions (e.g., π^2). *For example, by truncating the decimal expansion of $\sqrt{2}$, show that $\sqrt{2}$ is between 1 and 2, then between 1.4 and 1.5, and explain how to continue on to get better approximations.*	2–4
Domain: Expressions and Equations		
Work with radicals and integer exponents.		
8.EE.1	Know and apply the properties of integer exponents to generate equivalent numerical expressions. *For example, $3^2 \times 3^{-5} = 3^{-3} = \frac{1}{3^3} = \frac{1}{27}$.*	5
8.EE.2	Use square root and cube root symbols to represent solutions to equations of the form $x^2 = p$ and $x^3 = p$, where p is a positive rational number. Evaluate square roots of small perfect squares and cube roots of small perfect cubes. Know that $\sqrt{2}$ is irrational.	2, 6
8.EE.3	Use numbers expressed in the form of a single digit times an integer power of 10 to estimate very large or very small quantities, and to express how many times as much one is than the other. *For example, estimate the population of the United States as 3×10^8 and the population of the world as 7×10^9, and determine that the world population is more than 20 times larger.*	8
8.EE.4	Perform operations with numbers expressed in scientific notation, including problems where both decimal and scientific notation are used. Use scientific notation and choose units of appropriate size for measurements of very large or very small quantities (e.g., use millimeters per year for seafloor spreading). Interpret scientific notation that has been generated by technology.	7, 8
Understand the connections between proportional relationships, lines, and linear equations.		
8.EE.5	Graph proportional relationships, interpreting the unit rate as the slope of the graph. Compare two different proportional relationships represented in different ways. *For example, compare a distance-time graph to a distance-time equation to determine which of two moving objects has greater speed.*	12–14, 23
8.EE.6	Use similar triangles to explain why the slope m is the same between any two distinct points on a non-vertical line in the coordinate plane; derive the equation $y = mx$ for a line through the origin and the equation $y = mx + b$ for a line intercepting the vertical axis at b.	11, 12

Common Core State Standard	Grade 8	Coach Lesson(s)
Analyze and solve linear equations and pairs of simultaneous linear equations.		
8.EE.7	Solve linear equations in one variable.	
8.EE.7.a	Give examples of linear equations in one variable with one solution, infinitely many solutions, or no solutions. Show which of these possibilities is the case by successively transforming the given equation into simpler forms, until an equivalent equation of the form $x = a$, $a = a$, or $a = b$ results (where a and b are different numbers).	9, 10
8.EE.7.b	Solve linear equations with rational number coefficients, including equations whose solutions require expanding expressions using the distributive property and collecting like terms.	9, 10
8.EE.8	Analyze and solve pairs of simultaneous linear equations.	15
8.EE.8.a	Understand that solutions to a system of two linear equations in two variables correspond to points of intersection of their graphs, because points of intersection satisfy both equations simultaneously.	16, 17
8.EE.8.b	Solve systems of two linear equations in two variables algebraically, and estimate solutions by graphing the equations. Solve simple cases by inspection. *For example, $3x + 2y = 5$ and $3x + 2y = 6$ have no solution because $3x + 2y$ cannot simultaneously be 5 and 6.*	16, 17
8.EE.8.c	Solve real-world and mathematical problems leading to two linear equations in two variables. *For example, given coordinates for two pairs of points, determine whether the line through the first pair of points intersects the line through the second pair.*	17, 18
Domain: Functions		
Define, evaluate, and compare functions.		
8.F.1	Understand that a function is a rule that assigns to each input exactly one output. The graph of a function is the set of ordered pairs consisting of an input and the corresponding output.	19
8.F.2	Compare properties of two functions each represented in a different way (algebraically, graphically, numerically in tables, or by verbal descriptions). *For example, given a linear function represented by a table of values and a linear function represented by an algebraic expression, determine which function has the greater rate of change.*	23
8.F.3	Interpret the equation $y = mx + b$ as defining a linear function, whose graph is a straight line; give examples of functions that are not linear. *For example, the function $A = s^2$ giving the area of a square as a function of its side length is not linear because its graph contains the points (1, 1), (2, 4) and (3, 9), which are not on a straight line.*	19, 20

Common Core State Standard	Grade 8	Coach Lesson(s)
Use functions to model relationships between quantities.		
8.F.4	Construct a function to model a linear relationship between two quantities. Determine the rate of change and initial value of the function from a description of a relationship or from two (x, y) values, including reading these from a table or from a graph. Interpret the rate of change and initial value of a linear function in terms of the situation it models, and in terms of its graph or a table of values.	20, 21
8.F.5	Describe qualitatively the functional relationship between two quantities by analyzing a graph (e.g., where the function is increasing or decreasing, linear or nonlinear). Sketch a graph that exhibits the qualitative features of a function that has been described verbally.	22
Domain: Geometry		
Understand congruence and similarity using physical models, transparencies, or geometry software.		
8.G.1	Verify experimentally the properties of rotations, reflections, and translations:	
8.G.1.a	Lines are taken to lines, and line segments to line segments of the same length.	24
8.G.1.b	Angles are taken to angles of the same measure.	24
8.G.1.c	Parallel lines are taken to parallel lines.	24
8.G.2	Understand that a two-dimensional figure is congruent to another if the second can be obtained from the first by a sequence of rotations, reflections, and translations; given two congruent figures, describe a sequence that exhibits the congruence between them.	24
8.G.3	Describe the effect of dilations, translations, rotations, and reflections on two-dimensional figures using coordinates.	24, 25
8.G.4	Understand that a two-dimensional figure is similar to another if the second can be obtained from the first by a sequence of rotations, reflections, translations, and dilations; given two similar two-dimensional figures, describe a sequence that exhibits the similarity between them.	25, 26
8.G.5	Use informal arguments to establish facts about the angle sum and exterior angles of triangles, about the angles created when parallel lines are cut by a transversal, and the angle-angle criterion for similarity of triangles. *For example, arrange three copies of the same triangle so that the sum of the three angles appears to form a line, and give an argument in terms of transversals why this is so.*	26–28
Understand and apply the Pythagorean Theorem.		
8.G.6	Explain a proof of the Pythagorean Theorem and its converse.	29
8.G.7	Apply the Pythagorean Theorem to determine unknown side lengths in right triangles in real-world and mathematical problems in two and three dimensions.	29, 31
8.G.8	Apply the Pythagorean Theorem to find the distance between two points in a coordinate system.	30, 31

Common Core State Standard	Grade 8	Coach Lesson(s)
Solve real-world and mathematical problems involving volume of cylinders, cones, and spheres.		
8.G.9	Know the formulas for the volumes of cones, cylinders, and spheres and use them to solve real-world and mathematical problems.	32
Domain: Statistics and Probability		
Investigate patterns of association in bivariate data.		
8.SP.1	Construct and interpret scatter plots for bivariate measurement data to investigate patterns of association between two quantities. Describe patterns such as clustering, outliers, positive or negative association, linear association, and nonlinear association.	33, 34
8.SP.2	Know that straight lines are widely used to model relationships between two quantitative variables. For scatter plots that suggest a linear association, informally fit a straight line, and informally assess the model fit by judging the closeness of the data points to the line.	34
8.SP.3	Use the equation of a linear model to solve problems in the context of bivariate measurement data, interpreting the slope and intercept. *For example, in a linear model for a biology experiment, interpret a slope of 1.5 cm/hr as meaning that an additional hour of sunlight each day is associated with an additional 1.5 cm in mature plant height.*	35
8.SP.4	Understand that patterns of association can also be seen in bivariate categorical data by displaying frequencies and relative frequencies in a two-way table. Construct and interpret a two-way table summarizing data on two categorical variables collected from the same subjects. Use relative frequencies calculated for rows or columns to describe possible association between the two variables. *For example, collect data from students in your class on whether or not they have a curfew on school nights and whether or not they have assigned chores at home. Is there evidence that those who have a curfew also tend to have chores?*	36

Domain 1 The Number System

Domain 1: Diagnostic Assessment for Lessons 1–4

Lesson 1   Rational Numbers
8.NS.1

Lesson 2   Irrational Numbers
8.NS.1, 8.NS.2, 8.EE.2

Lesson 3   Compare and Order
Rational and Irrational
Numbers
8.NS.1, 8.NS.2

Lesson 4   Estimate the Value of
Expressions
8.NS.2

Domain 1: Cumulative Assessment for Lessons 1–4

Domain 1: Diagnostic Assessment for Lessons 1–4

1. Which decimal is equivalent to $7\frac{3}{8}$?

    A. $7.3\overline{8}$

    B. 7.38

    C. 7.375

    D. 0.375

2. Which symbol makes this sentence true?

    $$\pi \bigcirc \sqrt{13}$$

    A. >

    B. <

    C. =

    D. +

3. The composition of Earth's atmosphere is about 20.95% oxygen, 78.084% nitrogen, and less than 1% other gases. Which shows the decimal equivalent of 20.95%?

    A. 0.02095

    B. 0.2095

    C. 2.095

    D. 20.95

4. Which point on the number line below best represents the value of 3π?

    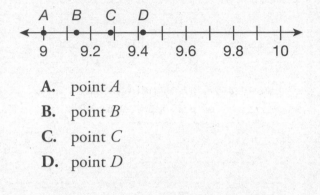

    A. point A

    B. point B

    C. point C

    D. point D

5. Which of the following is a rational number that can be written as a finite decimal?

    A. $\sqrt{3}$

    B. $\frac{2}{3}$

    C. 24.5%

    D. π

6. Consider the three irrational numbers below.

    $$2.828427\ldots, \pi, \sqrt{7}$$

    Which lists these numbers in order from least to greatest?

    A. $\pi, \sqrt{7}, 2.828427\ldots$

    B. $2.828427\ldots, \pi, \sqrt{7}$

    C. $\sqrt{7}, \pi, 2.828427\ldots$

    D. $\sqrt{7}, 2.828427\ldots, \pi$

7. Which point on the number line best represents $-1\frac{6}{7}$?

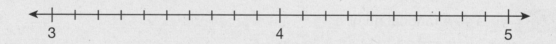

J   K   L   M

-2 -1.8 -1.6 -1.4 -1.2 -1

    A. point J

    B. point K

    C. point L

    D. point M

8. Which is closest to the value of $\sqrt{95} - \sqrt{18}$?

    A. 9.1

    B. 8.9

    C. 5.5

    D. 2.1

9. Estimate the value of $2\sqrt{17}$.

_____

10. Consider the number line below.

3   4   5

    A. Explain why $\sqrt{2}$ is an irrational number.

_____

_____

_____

    B. Estimate the value of $\sqrt{2} + \pi$ to the nearest hundredth. Show your work. Then graph a point to represent the value on the number line above.

Common Core State Standard:
8.NS.1

Rational Numbers

Getting the Idea

Integers include the set of **whole numbers** (0, 1, 2, 3, …) and their opposites (−1, −2, −3, …). The number line below shows integers from −5 to 5. Notice that positive numbers are located to the right of zero, and negative numbers are located to the left of zero.

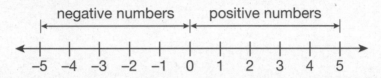

A **rational number** is any real number that can be expressed as the ratio of two integers $\frac{a}{b}$, where b is not equal to zero. Some examples of rational numbers are shown below.

$$-6, \; -\frac{3}{5}, \; 2\frac{7}{9}, \; 16\%, \; \sqrt{4}, \; 0.\overline{7}, \; 0.79$$

Any rational number can be expanded to form a decimal with digits that either terminate or repeat. To convert a fraction to a **repeating decimal**, use long division and divide the numerator by the denominator.

Example 1

Is $\frac{5}{11}$ a rational number? If so, write it as a decimal.

Strategy   **Decide if $\frac{5}{11}$ is rational. Then divide to write it as a decimal.**

> **Step 1**   Is $\frac{5}{11}$ a rational number?
>
> 5 and 11 are both integers.
>
> So, $\frac{5}{11}$ shows the ratio of two integers. It is rational.

> **Step 2**   Divide the numerator, 5, by the denominator, 11.
>
> Insert zeros after the decimal point in 5 as needed to perform the long division.

$$
\begin{array}{r}
0.4545\ldots \\
11\overline{)5.0000} \\
-\underline{4\,4} \\
60 \\
-\underline{55} \\
50 \\
-\underline{44} \\
60 \\
-\underline{55} \\
5
\end{array}
$$

Step 3   Write 0.4545… using a bar to show the repeating digits.

0.4545… = $0.\overline{45}$

Solution $\frac{5}{11}$ **is rational. It can be expressed as the repeating decimal $0.\overline{45}$.**

Some rational numbers can be expanded to form **finite decimals**. The digits of a finite decimal terminate and do not repeat. If you perform long division and the digit 0 keeps repeating, the decimal is finite and the zeros can be dropped.

Example 2

Is $-2\frac{3}{5}$ a rational number? If so, write it as a decimal.

Strategy   **Decide if $-2\frac{3}{5}$ is rational. Then divide to write it as a decimal.**

Step 1   Is $-2\frac{3}{5}$ a rational number?

Convert $-2\frac{3}{5}$ to an improper fraction.

$$-2\frac{3}{5} = -\frac{(2 \cdot 5) + 3}{5} = -\frac{10 + 3}{5} = -\frac{13}{5} = \frac{-13}{5}$$

Since $\frac{-13}{5}$ is the ratio of two integers, $-2\frac{3}{5}$ is rational.

Step 2   Divide the numerator, -13, by the denominator, 5.

Since you are dividing a negative integer by a positive integer, the quotient will be negative.

For now, drop the negative sign.

```
     2.600…
5)13.000
  −10
    3 0
   −3 0
     00
    −00
     00
    −00
     0
```

The actual quotient is $-2.600…$

Step 3   Write $-2.600…$ as a finite decimal.

Since $-2.600…$ ends in a sequence of zeros, the zeros can be dropped.

The quotient can be written as -2.6.

Solution $-2\frac{3}{5}$ **is rational. It can be expressed as the finite decimal -2.6.**

All rational numbers can be represented on a number line. To plot rational numbers on a number line, it is helpful to convert them to the same form.

You already know that you can convert a fraction to a decimal using long division. You can also convert a percent to a decimal by dividing the percent by 100 and dropping the percent sign. This is the same as moving the decimal point in the percent two places to the left.

Some square roots are also rational. Any number that has a whole-number **square root** is a **perfect square**, such as the ones below.

$$1, 4, 9, 16, 25, 36, 49, 64, 81, 100, 121, 144, 169, 196, 225$$

If the number under a square root symbol ($\sqrt{}$) is a perfect square, its value is an integer and it is a rational number. For example, $\sqrt{9}$ is rational because it is equal to the integer 3.

Example 3

Plot and label a point for each rational number below on a number line.

$$-\frac{2}{2}, 1\frac{1}{2}, -0.25, 72.5\%, \sqrt{4}$$

Strategy **Write the numbers in an equivalent form.**

Step 1 Rewrite each number as a decimal or integer.

$-\frac{2}{2} = -2 \div 2 = -1$

$\frac{1}{2} = 1 \div 2 = 0.5$, so $1\frac{1}{2} = 1 + 0.5 = 1.5$.

-0.25 is already in decimal form.

$72.5\% = 72.5\% \div 100\% = {}_{\wedge}72.5 = 0.725$

$\sqrt{4} = 2$, because $2^2 = 4$.

Step 2 Plot and label each number on a number line.

Draw a number line from -1 to 2 and divide it into tenths.

Plot a point at -1 and label it $-\frac{2}{2}$.

Plot a point for 1.5 at the tick mark halfway between 1.4 and 1.6 and label it $1\frac{1}{2}$.

Plot a point halfway between -0.2 and -0.3 and label it -0.25.

Plot a point for 0.725 closer to 0.7 than to 0.8 and label it 72.5%.

Plot a point at 2 and label it $\sqrt{4}$.

Solution **The number line with the rational numbers labeled is shown in Step 2.**

Coached Example

What decimal is represented by point *P* on the number line below?

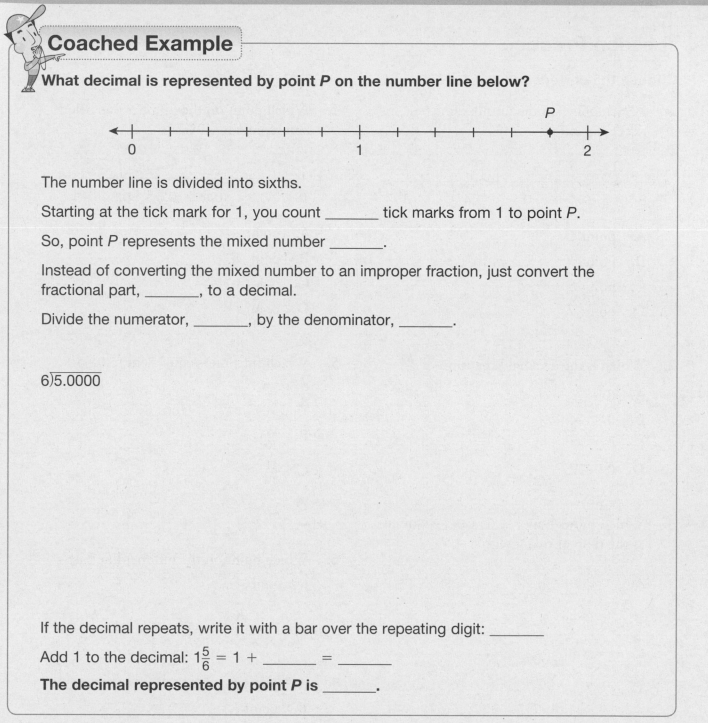

The number line is divided into sixths.

Starting at the tick mark for 1, you count _____ tick marks from 1 to point *P*.

So, point *P* represents the mixed number _____.

Instead of converting the mixed number to an improper fraction, just convert the fractional part, _____, to a decimal.

Divide the numerator, _____, by the denominator, _____.

$$6\overline{)5.0000}$$

If the decimal repeats, write it with a bar over the repeating digit: _____

Add 1 to the decimal: $1\frac{5}{6} = 1 +$ _____ $=$ _____

The decimal represented by point *P* is _____.

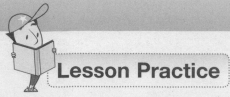

Lesson Practice

Choose the correct answer.

1. Which point on the number line best represents $-\frac{6}{2}$?

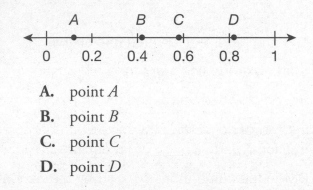

   A. point Q

   B. point R

   C. point S

   D. point T

2. Which is the decimal expansion of $\frac{3}{11}$?

   A. $0.\overline{34}$

   B. 0.3434

   C. $0.\overline{27}$

   D. 0.02727

3. Which shows how -2.65 can be written as the ratio of two integers?

   A. $-\frac{265}{20}$

   B. $-\frac{213}{20}$

   C. $-\frac{53}{20}$

   D. $-\frac{2}{25}$

4. Which point on the number line below best represents 42%?

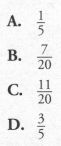

   A. point A

   B. point B

   C. point C

   D. point D

5. Which fraction is equivalent to 0.35?

   A. $\frac{1}{5}$

   B. $\frac{7}{20}$

   C. $\frac{11}{20}$

   D. $\frac{3}{5}$

6. Which point on the number line best represents $-2\frac{2}{3}$?

   A. point J

   B. point K

   C. point L

   D. point M

7. A gymnast is $4\frac{5}{12}$ feet tall. Which decimal is equivalent to $4\frac{5}{12}$?

   A. 4.416

   B. $4.41\overline{6}$

   C. 4.512

   D. $4.51\overline{2}$

8. The metal composition of a penny is 97.5% zinc and only 2.5% copper. How would 2.5% be written as a decimal?

   A. 2.500

   B. 2.05

   C. 0.25

   D. 0.025

9. Consider the number line below.

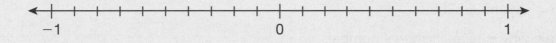

A. Write 65% as a decimal. Show your work. Then plot and label a point for it on the number line.

B. Write $-\frac{2}{9}$ as a decimal. Show your work. Then plot and label a point for it on the number line.

Common Core State Standards:
8.NS.1, 8.NS.2, 8.EE.2

Irrational Numbers

Getting the Idea

The set of **real numbers** includes both rational numbers and **irrational numbers**. Any real number that cannot be written as the ratio of two integers $\frac{a}{b}$, where b is not equal to zero, is irrational. An irrational number can be expressed as an infinite, non-repeating decimal.

Example 1

The value of the number pi (π) is shown below.

$\pi = 3.1459265358\ldots$

Explain why π is an irrational number.

Strategy   **Use the definition of an irrational number.**

Step 1   Recall the characteristics of irrational numbers.

The value of an irrational number is an infinite, non-repeating decimal.

Step 2   Describe the decimal value of pi.

The ellipsis (…) shows that the digits continue on forever and do not repeat.

So, π is an infinite, non-repeating decimal.

Solution   **Pi (π) is an irrational number because it is an infinite, non-repeating decimal.**

Pi is not the only irrational number. The square roots of positive numbers that are not perfect squares are also irrational. You can estimate the value of a square root by deciding which two perfect squares it lies between and then using guess and check to estimate its value more precisely.

Example 2

Explain why $\sqrt{2}$ is irrational. Then estimate its value.

Strategy   **Use the definition of an irrational number.**
Then use guess and check to estimate the value of $\sqrt{2}$.

Step 1   Explain why $\sqrt{2}$ is irrational.

There is no whole number that can be multiplied by itself to get 2.

So, $\sqrt{2}$ is irrational.

Step 2   Find which perfect squares the number 2 lies between.

You can find perfect squares by squaring consecutive whole numbers.

$1^2 = 1 \cdot 1 = 1$

$2^2 = 2 \cdot 2 = 4$

2 is between the perfect squares 1 and 4, so:

$\sqrt{1} < \sqrt{2} < \sqrt{4}$.

(The symbol $<$ means "is less than.")

Step 3   Find which consecutive whole numbers $\sqrt{2}$ lies between.

$\sqrt{1} = 1$, and $\sqrt{4} = 2$.

So, $1 < \sqrt{2} < 2$.

Step 4   Determine if the value is closer to 1 or 2.

2 is closer in value to 1 than to 4.

So, $\sqrt{2}$ is a little closer to 1 than to 2.

Step 5   Use guess and check to estimate $\sqrt{2}$ to the nearest tenth.

Try 1.4.

$1.4^2 = 1.4 \cdot 1.4 = 1.96$ → close, but slightly less than 2.

Try the next higher number, 1.5.

$1.5^2 = 1.5 \cdot 1.5 = 2.25$ → close, but not as close as 1.96.

Since 2 is closer to 1.4^2 than to 1.5^2, $\sqrt{2} \approx 1.4$.

(The symbol $\approx$ means "is approximately equal to.")

Note: If you wanted to get an even better estimate, you could use guess and check to find that 1.41^2 is closer to 2 than is 1.42^2, and so on.

Solution   **The number $\sqrt{2}$ is irrational because 2 is not a perfect square.**
Its value is an infinite, non-repeating decimal close to 1.4.

If you have a calculator, you can use it to check your answer for Example 2. The calculator will show that $\sqrt{2} = 1.414213562...$, which is an infinite, non-repeating decimal.

Example 3

Graph the approximate location of $\sqrt{34}$ on a number line.

Strategy   **Use guess and check to estimate the value of $\sqrt{34}$ to the nearest tenth. Then graph the decimal on a number line.**

Step 1   Find which perfect squares 34 lies between.

$$5^2 = 5 \cdot 5 = 25$$

$$6^2 = 6 \cdot 6 = 36$$

34 is between the perfect squares 25 and 36, so: $\sqrt{25} < \sqrt{34} < \sqrt{36}$.

Step 2   Find which consecutive whole numbers $\sqrt{34}$ lies between.

$\sqrt{25} = 5$, and $\sqrt{36} = 6$, so: $5 < \sqrt{34} < 6$.

34 is closer to 36 than to 25, so $\sqrt{34}$ is closer to 6 than to 5.

Step 3   Use guess and check to estimate $\sqrt{34}$ to the nearest tenth.

Try 5.9.

$5.9^2 = 5.9 \cdot 5.9 = 34.81$ ➞   close, but more than 34.

Try the next lower number, 5.8.

$5.8^2 = 5.8 \cdot 5.8 = 33.64$ ➞   close, and less than 34.

Since 34 is closer to 5.8^2 than to 5.9^2, $\sqrt{34} \approx 5.8$.

Step 4   Graph $\sqrt{34}$ on a number line divided into tenths.

Plot $\sqrt{34}$ between 5.8 and 5.9, but closer to 5.8.

$$\sqrt{34}$$

```
   ◄──┼────┼────┼────┼────┼────┼────┼────┼────●──┼────┼──►
      5   5.1  5.2  5.3  5.4  5.5  5.6  5.7  5.8  5.9   6
```

Solution   **$\sqrt{34}$ is graphed on the number line in Step 4.**

Coached Example

The area of a square is 67 square meters. Find the exact length, in meters, of one side of the square. Then graph that value on a number line.

The area, A, of a square is found using the formula $A = s^2$, where s shows the length of one side.

So, the length of one side, s, can be found by taking the square root of _____.

The exact length of each side of the square is $\sqrt{\rule{2cm}{0pt}}$ meters.

To graph that number on a number line, first estimate its value as a decimal.

67 lies between the perfect squares 64 and _____.

$\sqrt{64}$ = _____, and the square root of the other perfect square is _____.

So, $\sqrt{67}$ lies between the whole numbers _____ and _____, but is closer to _____.

Use guess and check to estimate its value to the nearest tenth.

Try 8.1:

    $8.1^2 = 8.1 \cdot 8.1 =$ _____   →   close, but _____ than 67.

Try 8.2:

    $8.2^2 = 8.2 \cdot 8.2 =$ _____   →   close, and _____ than 67.

Which is closer to 67: 8.1^2 or 8.2^2? _____

So, $\sqrt{67}$ is between 8.1 and 8.2, but is closer to _____.

Graph $\sqrt{67}$ on the number line below.

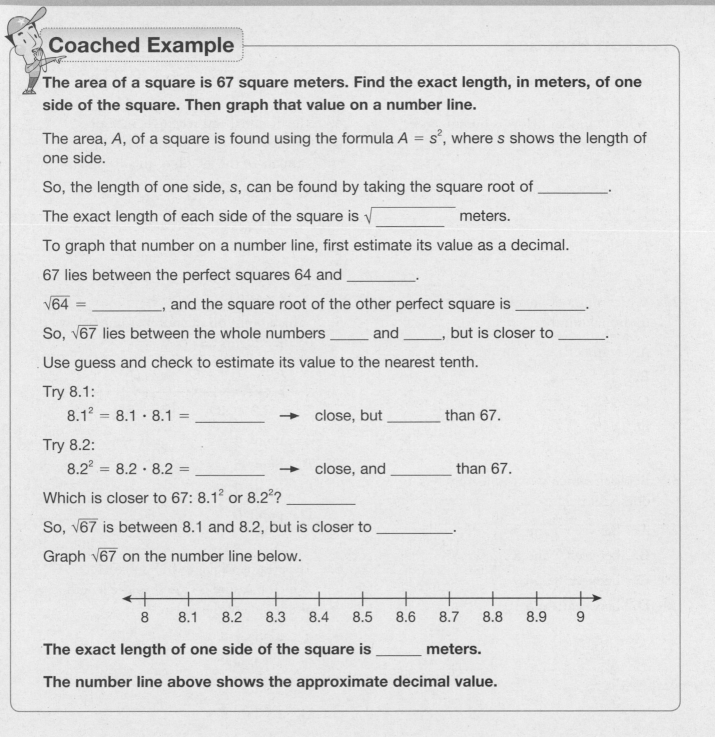

The exact length of one side of the square is _____ meters.

The number line above shows the approximate decimal value.

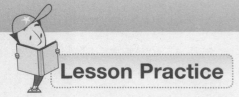

Lesson Practice

Choose the correct answer.

1. Which number below is **not** an irrational number?

    A. π

    B. $\sqrt{2}$

    C. $\sqrt{4}$

    D. $\sqrt{6}$

2. Which number below is **not** an irrational number?

    A. $\sqrt{46}$

    B. $\sqrt{47}$

    C. $\sqrt{48}$

    D. $\sqrt{49}$

3. Between which two whole numbers does $\sqrt{89}$ lie?

    A. between 4 and 5

    B. between 7 and 8

    C. between 8 and 9

    D. between 9 and 10

4. The diagonal of a rectangle measures $\sqrt{5}$ meters long. Which is the best estimate of the length of the diagonal?

    A. 1.9 meters

    B. 2.2 meters

    C. 2.3 meters

    D. 2.5 meters

5. Which point on the number line below best represents $\sqrt{11}$?

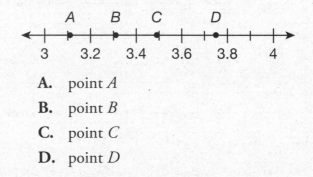

    A. point A

    B. point B

    C. point C

    D. point D

6. The area of a square is 17 square feet. Which measure is closest to the length of one side of the square?

    A. 4.2 feet

    B. 4.1 feet

    C. 3.9 feet

    D. 1.7 feet

7. Which point on the number line below best represents $\sqrt{50}$?

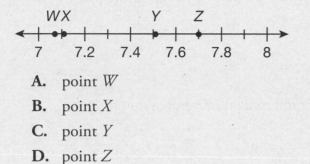

A. point W

B. point X

C. point Y

D. point Z

8. One side of a triangular garden is $\sqrt{15}$ feet long. Which is the best estimate of the length of the side?

A. 3.17 feet

B. 3.18 feet

C. 3.87 feet

D. 3.92 feet

9. Consider the number line below.

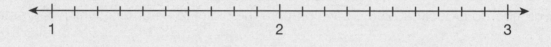

A. Explain why $\sqrt{3}$ is an irrational number.

_____

_____

_____

B. Estimate the value of $\sqrt{3}$ to the nearest tenth. Show your work. Then graph it on the number line above.

Compare and Order Rational and Irrational Numbers

Common Core State Standards:
8.NS.1, 8.NS.2

Getting the Idea

Sometimes, you may need to compare or order rational and irrational numbers.

The symbols below will help you do this:

> means "is greater than."

< means "is less than."

= means "is equal to."

To compare an irrational number to another number, estimate its decimal value. Convert the other number to a decimal also. Then compare the digits to determine which is greater.

Example 1

Which symbol (>, <, or =) makes this sentence true?

$$7.745966\ldots \bigcirc \sqrt{59}$$

Strategy   **Estimate the values of the irrational numbers.**

Step 1   Round 7.745966... to the nearest hundredth.

$$7.745966\ldots \approx 7.75$$

Step 2   Estimate $\sqrt{59}$ to the nearest whole number.

$\sqrt{49} < \sqrt{59} < \sqrt{64}$, so:

$$7 < \sqrt{59} < 8.$$

59 is closer to 64 than to 49, so $\sqrt{59}$ is closer to 8.

Step 3   Continue estimating and compare.

$7.7^2 = 7.7 \cdot 7.7 = 59.29 \quad \longrightarrow \quad$ close, but more than 59

You do not need to estimate further.

Since 59 is less than 7.7^2, $\sqrt{59}$ must be less than 7.7.

$7.75 > 7.7$, so $7.745966\ldots > \sqrt{59}$.

Solution   **The symbol > makes the sentence true.**

$$7.745966\ldots \bigcirc\!\!\!> \sqrt{59}$$

When ordering a set of numbers with different signs, remember that a positive number is always greater than a negative number.

Example 2

Order the numbers below from least to greatest.

$28\%, -2\frac{1}{2}, \frac{2}{7}, -\sqrt{9}$

Strategy   **Separate the negative numbers from the positive numbers. Then convert each group of numbers to the same form.**

Step 1   Write the negative numbers as decimals and compare.

$-2\frac{1}{2} = -(2 + \frac{1}{2}) = -(2 + 0.5) = -2.5$

$-\sqrt{9} = -(3) = -3$

Sketching a number line shows that $-3 < -2.5$, so $-\sqrt{9} < -2.5$.

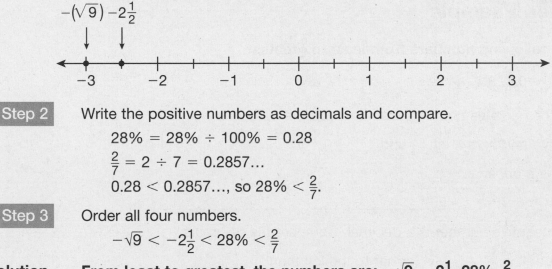

Step 2   Write the positive numbers as decimals and compare.

$28\% = 28\% \div 100\% = 0.28$

$\frac{2}{7} = 2 \div 7 = 0.2857...$

$0.28 < 0.2857...$, so $28\% < \frac{2}{7}$.

Step 3   Order all four numbers.

$-\sqrt{9} < -2\frac{1}{2} < 28\% < \frac{2}{7}$

Solution   **From least to greatest, the numbers are: $-\sqrt{9}, -2\frac{1}{2}, 28\%, \frac{2}{7}$.**

Example 3

Order these numbers from greatest to least.

$\pi, 3\frac{1}{3}, \sqrt{14}$

Strategy   **Estimate the value of each number.**

Step 1   Approximate the values of π and $3\frac{1}{3}$.

$\pi \approx 3.14$

$3\frac{1}{3} = 3 + \frac{1}{3} = 3 + 0.333... \approx 3.33$

Estimate $\sqrt{14}$ to the nearest tenth.

$\sqrt{9} < \sqrt{14} < \sqrt{16}$, so:

$3 < \sqrt{14} < 4$.

14 is closer to 16 than to 9, so $\sqrt{14}$ is closer to 4.

$3.7^2 = 3.7 \cdot 3.7 = 13.69$ → close

$3.8^2 = 3.8 \cdot 3.8 = 14.44$ → not as close as 3.7^2

$\sqrt{14} \approx 3.7$

Step 3 Order the decimals and then the numbers.

$3.7 > 3.33 > 3.14$

$\sqrt{14} > 3\frac{1}{3} > \pi$

Solution **From greatest to least, the order of the numbers is: $\sqrt{14}$, $3\frac{1}{3}$, π.**

Coached Example

Order the following numbers from least to greatest.

$-\frac{1}{9}$, $\sqrt{5}$, -0.8, 3.5

Separate the negative numbers from the positive numbers.

The negative numbers are: $-\frac{1}{9}$ and _____.

Write $-\frac{1}{9}$ as a decimal:

$-\frac{1}{9} = -1 \div 9 = $ _____

The other negative number is a decimal. Compare the decimals.

On a number line, _____ is farther to the left than _____.

So, _____ < _____.

The positive numbers are: $\sqrt{5}$ and _____.

Estimate $\sqrt{5}$ to the nearest tenth.

$\sqrt{4} < \sqrt{5} < \sqrt{9}$, so:

$2 < \sqrt{5} < $ _____.

Since the value of $\sqrt{5}$ is less than 3, 3.5 must be _____ than $\sqrt{5}$.

From least to greatest, the order is: _____, _____, _____, _____.

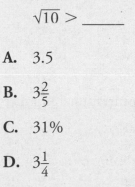

Lesson Practice

Choose the correct answer.

1. Which symbol makes this sentence true?

   $-\sqrt{7} \bigcirc -3.5$

   A. >

   B. <

   C. =

   D. +

2. Which symbol makes this sentence true?

   $\sqrt{2} \bigcirc 1.7320508...$

   A. >

   B. <

   C. =

   D. +

3. Which symbol makes this sentence true?

   $\pi \bigcirc \frac{28}{9}$

   A. >

   B. <

   C. =

   D. +

4. Which number goes in the blank to make this sentence true?

   $\sqrt{10} > $ _____

   A. 3.5

   B. $3\frac{2}{5}$

   C. 31%

   D. $3\frac{1}{4}$

5. Which set of numbers is ordered from least to greatest?

   A. 27%, 2.75, $\sqrt{2}$, $2\frac{7}{9}$

   B. 27%, $\sqrt{2}$, 2.75, $2\frac{7}{9}$

   C. $\sqrt{2}$, $2\frac{7}{9}$, 2.75, 27%

   D. $2\frac{7}{9}$, 2.75, $\sqrt{2}$, 27%

6. Which set of numbers is ordered from greatest to least?

   A. π, $32\frac{2}{3}$%, $\frac{7}{3}$, $\sqrt{6}$

   B. π, $\sqrt{6}$, $\frac{7}{3}$, $32\frac{2}{3}$%

   C. $\sqrt{6}$, $32\frac{2}{3}$%, $\frac{7}{3}$, π

   D. $\frac{7}{3}$, π, $32\frac{2}{3}$%, $\sqrt{6}$

7. Which set of numbers is ordered from least to greatest?

   A. $\frac{9}{11}$, -0.9, 95%, $-\sqrt{1}$

   B. -0.9, $-\sqrt{1}$, $\frac{9}{11}$, 95%

   C. $-\sqrt{1}$, -0.9, 95%, $\frac{9}{11}$

   D. $-\sqrt{1}$, -0.9, $\frac{9}{11}$, 95%

8. Which number goes in the blank to make this sentence true?

   $$\frac{5}{6} < \underline{\hspace{1cm}} < 85\%$$

   A. π

   B. $\sqrt{64}$

   C. 0.84

   D. 83%

9. Consider the three irrational numbers below.

   $\sqrt{11}$, π, $3.741657\ldots$

   A. Estimate the value of each number as a decimal to the nearest tenth, showing each step in the process.

   B. Order the original numbers from least to greatest. Explain your reasoning.

   _____

   _____

Estimate the Value of Expressions

Common Core State Standard:
8.NS.2

Getting the Idea

Sometimes, you may need to estimate the value of an **expression** that includes an irrational number. To do that, estimate the value of the irrational number. Then use that estimate to find the value of the entire expression.

When estimating the value of an expression that includes π, it is helpful to remember that π can be approximated as 3.14 or $\frac{22}{7}$.

Example 1

Estimate the value of π^2.

Strategy   **Substitute 3.14 for π.**

   Step 1   Write an expression that could be used.

   The exponent, 2, indicates that π is used as a factor two times.

   $\pi^2 = \pi \cdot \pi \approx 3.14 \cdot 3.14$

   Step 2   Multiply.

   Since each factor has 2 decimal places, the product will have $2 + 2$, or 4, decimal places.

   $3.14 \cdot 3.14 = 9.8596 \approx 9.86$

   NOTE: If you have a calculator, use it to check the answer:

   $\pi^2 = 9.86960440\ldots$ ✓ This is close to 9.86.

Solution   **The value of π^2 can be estimated as 9.86.**

Example 2

Estimate the value of $2\sqrt{7}$.

Strategy   **Estimate the value of $\sqrt{7}$ to the nearest tenth.**

Step 1   Estimate $\sqrt{7}$ to the nearest whole number.

$\sqrt{4} < \sqrt{7} < \sqrt{9}$, so $2 < \sqrt{7} < 3$.

7 is slightly closer to 9 than to 4, so $\sqrt{7}$ is slightly closer to 3.

Step 2   Use guess and check to estimate $\sqrt{7}$ to the nearest tenth.

$2.6^2 = 2.6 \cdot 2.6 = 6.76 \longrightarrow$ close

$2.7^2 = 2.7 \cdot 2.7 = 7.29 \longrightarrow$ not as close as 2.6^2

$\sqrt{7} \approx 2.6$

Step 3   Use that decimal approximation to estimate the value of the expression.

$2\sqrt{7} = 2 \cdot \sqrt{7} \approx 2 \cdot 2.6 \approx 5.2$

Solution   **A good estimate of the value of $2\sqrt{7}$ is 5.2.**

Example 3

Estimate the value of this expression.

$\sqrt{43} - \frac{7}{11}$

Strategy   **Estimate the value of each number. Then subtract.**

Step 1   Estimate $\sqrt{43}$ to the nearest whole number.

$\sqrt{36} < \sqrt{43} < \sqrt{49}$, so $6 < \sqrt{43} < 7$.

43 is slightly closer to 49 than to 36, so $\sqrt{43}$ is slightly closer to 7.

Step 2   Use guess and check to estimate $\sqrt{43}$ to the nearest tenth.

$6.5^2 = 6.5 \cdot 6.5 = 42.25 \longrightarrow$ close

$6.6^2 = 6.6 \cdot 6.6 = 43.56 \longrightarrow$ closer than 6.5^2

$\sqrt{43} \approx 6.6$

Step 3   Estimate $\frac{7}{11}$ to the nearest tenth.

$\frac{7}{11} = 7 \div 11 = 0.6363\ldots \approx 0.6$

Step 4   Subtract the estimated values of the numbers.

$\sqrt{43} - \frac{7}{11} \approx 6.6 - 0.6 \approx 6.0$

Solution   **The value of $\sqrt{43} - \frac{7}{11}$ is approximately 6.**

Coached Example

Estimate the value of $\frac{\pi}{2}$. Plot a point to represent that value on a number line.

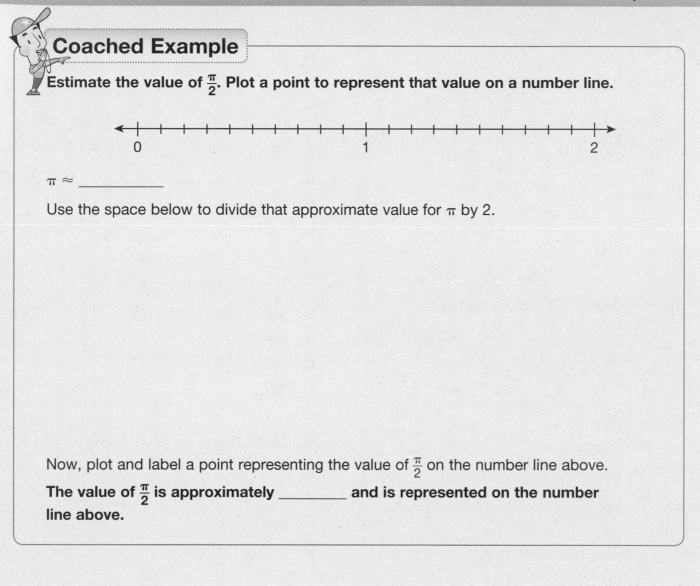

$\pi \approx$ _____

Use the space below to divide that approximate value for π by 2.

Now, plot and label a point representing the value of $\frac{\pi}{2}$ on the number line above.

The value of $\frac{\pi}{2}$ is approximately _____ and is represented on the number line above.

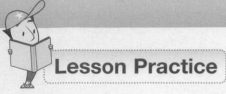

Lesson Practice

Choose the correct answer.

1. Which is the best estimate of the value of $3\sqrt{8}$?

   A. 2.4

   B. 6.3

   C. 7.9

   D. 8.4

2. Which point on the number line is closest to the value of 2π?

   ```
        A   B C          D
   ←──┼─●─┼─●●─┼───┼───┼─●─┼───┼→
      6   6.2   6.4   6.6   6.8   7
   ```

   A. point A

   B. point B

   C. point C

   D. point D

3. Which is the best estimate of the value of $\frac{11}{3} + \sqrt{11}$?

   A. 6.0

   B. 7.0

   C. 7.3

   D. 14.7

4. Which is closest to the value of $5\sqrt{15}$?

   A. 19.5

   B. 19.0

   C. 8.8

   D. 4.5

5. Which is closest to the value of $\frac{\sqrt{78}}{4}$?

   A. 4.8

   B. 2.4

   C. 2.2

   D. 1.95

6. Which is closest to the value of $\sqrt{90} - \frac{4}{11}$?

   A. 7.8

   B. 8.9

   C. 9.1

   D. 9.5

7. Which is closest to the value of $\frac{\sqrt{19}}{\sqrt{5}}$?

    A. 2.0

    B. 3.8

    C. 4.0

    D. 9.5

8. Which is the best estimate of the value of π^3?

    A. 9.4

    B. 27.8

    C. 31.0

    D. 97.2

9. The expression $\pi - \sqrt{20}$ contains two irrational numbers.

    A. Estimate the value of the expression $\pi - \sqrt{20}$. Show your work.

    B. Graph a point to represent the approximate value of $\pi - \sqrt{20}$ on the number line below.

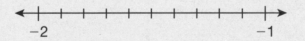

Domain 1: Cumulative Assessment for Lessons 1–4

1. The number -4.2 is rational. Which shows that number expressed as the ratio of two integers?

   A. $-\frac{4}{2}$

   B. $-\frac{12}{5}$

   C. $-\frac{21}{5}$

   D. $-\frac{40}{2}$

2. Which symbol makes this sentence true?

   $$\pi \bigcirc \sqrt{7}$$

   A. $>$

   B. $<$

   C. $=$

   D. $+$

3. Oceans cover approximately 70.8% of Earth's surface. Which shows 70.8% expressed as a decimal?

   A. 0.0708

   B. 0.708

   C. 0.78

   D. 7.08

4. Consider the three irrational numbers below.

   $$3.24037\ldots,\ \pi,\ \sqrt{12}$$

   Which lists these numbers in order from greatest to least?

   A. $\pi,\ \sqrt{12},\ 3.24037\ldots$

   B. $\pi,\ 3.24037\ldots,\ \sqrt{12}$

   C. $3.24037\ldots,\ \pi,\ \sqrt{12}$

   D. $\sqrt{12},\ 3.24037\ldots,\ \pi$

5. Which of the following is a rational number that can be written as a decimal in which one or more nonzero digits repeat?

   A. $\sqrt{16}$

   B. $\sqrt{5}$

   C. $\frac{5}{8}$

   D. $\frac{5}{9}$

6. Which best represents the value of $2\sqrt{45}$?

   A. 13.6

   B. 13.4

   C. 9.0

   D. 8.7

7. Which point on the number line best represents $2\frac{2}{11}$?

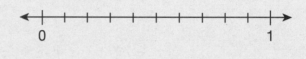

A. point W

B. point X

C. point Y

D. point Z

8. Which is closest to the value of $\sqrt{24} + \sqrt{18}$?

A. 9.1

B. 8.9

C. 4.2

D. 2.1

9. On the number line below, plot and label a point that represents the approximate value of $\frac{\pi}{10}$.

10. Consider the statement below.

$\sqrt{2}$ ◯ 1.5811388…

A. Explain why $\sqrt{2}$ and 1.5811388… are irrational numbers.

_____

_____

_____

_____

B. Estimate the value of $\sqrt{2}$ to the nearest tenth. Then choose the symbol ($<$, $>$, or $=$) that makes the sentence above true. Show your work and explain how you determined your answer.

Domain 2 Expressions and Equations

Domain 2: Diagnostic Assessment for Lessons 5–18

Lesson 5 Exponents
8.EE.1

Lesson 6 Square Roots and
Cube Roots
8.EE.2

Lesson 7 Scientific Notation
8.EE.4

Lesson 8 Solve Problems Using
Scientific Notation
8.EE.3, 8.EE.4

Lesson 9 Linear Equations in
One Variable
8.EE.7.a, 8.EE.7.b

Lesson 10 Use One-Variable Linear
Equations to Solve
Problems
8.EE.7.a, 8.EE.7.b

Lesson 11 Slope
8.EE.6

Lesson 12 Slopes and *y*-intercepts
8.EE.5, 8.EE.6

Lesson 13 Proportional Relationships
8.EE.5

Lesson 14 Direct Proportions
8.EE.5

Lesson 15 Pairs of Linear Equations
8.EE.8

Lesson 16 Solve Systems of
Equations Graphically
8.EE.8.a, 8.EE.8.b

Lesson 17 Solve Systems of
Equations Algebraically
8.EE.8.a, 8.EE.8.b, 8.EE.8.c

Lesson 18 Use Systems of Equations
to Solve Problems
8.EE.8.c

Domain 2: Cumulative Assessment for Lessons 5–18

Domain 2: Diagnostic Assessment for Lessons 5–18

1. Which shows $8^{-3} \times 8^5$ in exponential form?

   A. 8^{-15}

   B. 8^{-8}

   C. 8^2

   D. 8^8

2. A half dollar has a diameter of 3×10^{-2} meter. If an atom has a diameter of 5×10^{-10} meter, about how many of those atoms could fit across the diameter of the coin?

   A. 6×10^8

   B. 6×10^7

   C. 1.7×10^{-8}

   D. 1.7×10^{-12}

3. Which best describes the solution for this equation?

$$-6p - 4 = -2(3p + 2)$$

   A. $p = -8$

   B. $p = 0$

   C. no solution

   D. infinitely many solutions

4. Emma is buying potatoes from a farm stand. The graph below shows the cost.

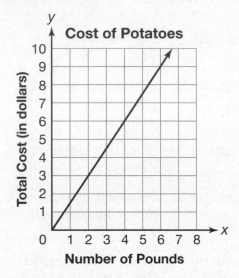

What does the slope of the graph represent?

   A. the unit price, $1.50 per pound of potatoes

   B. the unit price, $3.00 per pound of potatoes

   C. the number of pounds Emma bought, 6 pounds of potatoes

   D. the total amount Emma spent, $9.00

5. Solve the system of linear equations.

$$4x + 3y = 4$$
$$4x + 3y = -8$$

   A. $(2, 4)$

   B. $(2, -4)$

   C. no solution

   D. infinitely many solutions

6. The graph below represents $y = -2x$.

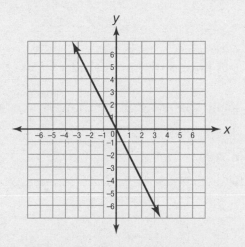

Which describes how this graph would need to be shifted in order to graph $y = -2x - 5$?

A. Shift each point 5 units up.

B. Shift each point 2 units up.

C. Shift each point 5 units down.

D. Shift each point 2 units down.

7. Which equation represents a line that is parallel, but not coincident, to $y = \frac{1}{2}x - 3$?

A. $y = \frac{2}{4}x - 6$

B. $y = \frac{3}{6}x - 3$

C. $y = 2x - 6$

D. $y = -\frac{1}{2}x - 1$

8. Two jets, Jet 1 and Jet 2, are cruising at a constant rate of speed. The equation $y = 575x$ shows the total distance, y, in miles, traveled by Jet 1 over x hours. The graph below shows the relationship between the traveling time and the distance traveled for Jet 2.

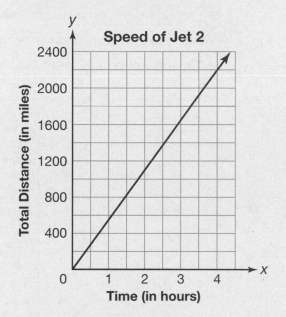

Which statement is true?

A. Jet 1 is traveling at a faster speed than Jet 2.

B. Jet 2 is traveling at a faster speed than Jet 1.

C. Jet 1 and Jet 2 are traveling at the same speed.

D. Jet 1 is sometimes traveling at a faster speed than Jet 2 and sometimes at a slower speed than Jet 2.

9. Solve for x.

$$x^2 = 36$$

10. Bella wants to paint some ceramics at a shop. The rates charged by two different shops are shown below.

Shop 1	Shop 2
$8 initial fee plus $2 per hour	$4 per hour only (no initial fee)

A. Let *x* represent the number of hours that Bella spends painting at the shop.

Let *y* represent the total cost of painting at the shop.

Write a system of equations to represent this problem.

Then graph the system on the coordinate grid below.

_____

_____

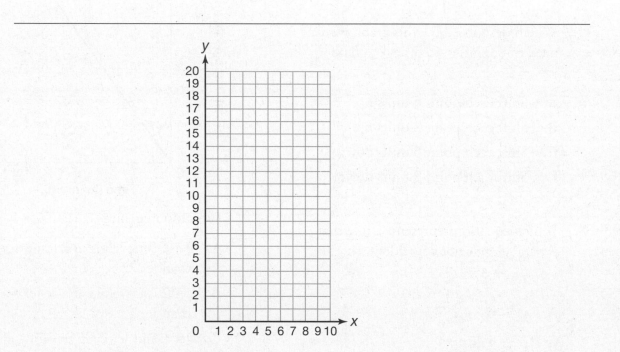

B. How many hours would Bella spend painting ceramics in order for the total cost to be the same at both shops? What would that total cost be?

Use the solution for the system of linear equations to explain your answer.

_____

_____

_____

Exponents

Common Core State Standard:
8.EE.1

Getting the Idea

A number in exponential form has a **base** and an **exponent**. The exponent indicates how many times the base is used as a factor. In a^s, the base is a and the exponent is s.

5^4 indicates that 5 is used as a factor 4 times: $5^4 = 5 \times 5 \times 5 \times 5 = 625$.

5^4 written in standard form is 625. 5^4 can also be called a power of 5 and is read as "five to the fourth power."

Note that the value of a nonzero expression in which the exponent is 0 is 1, so $5^0 = 1$.

Example 1

What is 4^3 written in standard form?

Strategy **Multiply the base by itself the number of times shown by the exponent.**

$$4 \times 4 \times 4 = 64$$

Solution $4^3 = 64$ **in standard form.**

A base raised to a negative exponent is equal to the **reciprocal** of the expression with a positive exponent. (Remember that the reciprocal of a is $\frac{1}{a}$.) Look at the examples below.

$$5^{-3} = \left(\frac{1}{5}\right)^3 = \frac{1}{5^3} \qquad \frac{1}{6^{-2}} = \left(\frac{6}{1}\right)^2 = 6^2$$

Reciprocals:

To change the sign of an exponent, move the expression to the denominator of a fraction.

$$a^{-n} = \frac{1}{a^n}, \text{ if } a \neq 0$$

To change the sign of an exponent in a denominator, move the expression to the numerator.

$$\frac{1}{a^{-n}} = \frac{a^n}{1} = a^n, \text{ if } a \neq 0$$

Example 2

What is 8^{-3} written in standard form?

Strategy   **Write the reciprocal of the exponential expression with a positive exponent, then simplify.**

Step 1   Write the reciprocal of the exponential expression to eliminate the negative exponent.

$$8^{-3} = \left(\frac{1}{8}\right)^3 = \frac{1}{8^3}$$

Step 2   Simplify.

$$\frac{1}{8^3} = \frac{1}{8 \times 8 \times 8} = \frac{1}{512}$$

Solution $8^{-3} = \dfrac{1}{512}$ **in standard form.**

Numbers in exponential form are sometimes called **powers**. There are some properties you can apply to simplify powers. In the table below, a and b are real numbers, and m and n are integers.

Properties of Powers	Examples
Product of Powers To multiply two numbers with the same base, add the exponents.	$a^m \times a^n = a^{m+n}$ $8^2 \times 8^7 = 8^{2+7} = 8^9$
Quotient of Powers To divide two numbers with the same base, subtract the exponents.	$a^m \div a^n = a^{m-n}$ $3^{10} \div 3^2 = 3^{10-2} = 3^8$
Power of a Power To raise a power to a power, multiply the exponents.	$(a^m)^n = a^{m \times n}$ $(6^4)^5 = 6^{4 \times 5} = 6^{20}$
Power of Zero Any nonzero number raised to the power of zero is 1.	$a^0 = 1$, if $a \neq 0$ $4^0 = 1$
Power of a Product To find a power of a product, find the power of each factor and multiply.	$(ab)^m = a^m b^m$ $(3 \times 2)^3 = 3^3 \times 2^3$
Power of a Quotient To raise a quotient to a power, raise both the numerator and denominator to that power.	$\left(\dfrac{a}{b}\right)^m = \dfrac{a^m}{b^m}$, if $b \neq 0$

Example 3

What is 25^0 written in standard form?

Strategy   **Use the power of zero property.**

$a^0 = 1$, if $a \neq 0$

$25 \neq 0$, so $25^0 = 1$.

Solution   **$25^0 = 1$ in standard form.**

Example 4

What is $3^2 \times 3^4$? Write the product in standard form.

Strategy   **Use the product of powers property.**

Step 1   The exponential terms have the same base, 3. Add the exponents.

$a^m \times a^n = a^{m+n}$

$3^2 \times 3^4 = 3^{2+4} = 3^6$

Step 2   Evaluate. Write the number in standard form.

$3^6 = 3 \times 3 \times 3 \times 3 \times 3 \times 3 = 729$

Solution   **$3^2 \times 3^4 = 3^6 = 729$**

Note: You could have solved the problem by evaluating each exponent and then multiplying. Since $3^2 = 9$ and $3^4 = 81$, $9 \times 81 = 729$.

Example 5

What is $5^5 \div 5^3$? Write the quotient in standard form.

Strategy   **Use the quotient of powers property.**

Step 1   The exponential terms have the same base, 5. Subtract the exponents.

$a^m \div a^n = a^{m-n}$

$5^5 \div 5^3 = 5^{5-3} = 5^2$

Step 2   Evaluate. Write the number in standard form.

$5^2 = 5 \times 5 = 25$

Solution   **$5^5 \div 5^3 = 5^2 = 25$**

Example 6

What is $7^3 \times 7^{-5}$? Write the product in standard form.

Strategy   **Use the product of powers property.**

Step 1   The exponential terms have the same base. Add the exponents.

$$a^m \times a^n = a^{m+n}$$
$$7^3 \times 7^{-5} = 7^{3+(-5)} = 7^{-2}$$

Step 2   Write the exponential expression without the negative exponent.

$$7^{-2} = \frac{1}{7^2}$$

Step 3   Evaluate. Write the number in standard form.

$$\frac{1}{7^2} = \frac{1}{7 \times 7} = \frac{1}{49}$$

Solution $7^3 \times 7^{-5} = 7^{-2} = \dfrac{1}{49}$

Coached Example

What is $(10^2)^3$ in standard form?

To raise a power to a power, you must _____ the exponents.

$$(10^2)^3 = 10^{2\;—\;3} = 10^{—}$$

Use 10 as a factor _____ times.

Multiply to find the product in standard form.

_____

$(10^2)^3 =$ _____

Lesson Practice

Choose the correct answer.

1. Which shows 5^4 in standard form?

   A. 20

   B. 625

   C. 1,024

   D. 3,125

2. Which is $6^3 \times 6^4$ in exponential form?

   A. 36^{12}

   B. 7^6

   C. 6^{12}

   D. 6^7

3. Which shows 9^{-3} in standard form?

   A. 729

   B. 27

   C. $\frac{1}{27}$

   D. $\frac{1}{729}$

4. Which shows $(11^6)^2$ in exponential form?

   A. 22^6

   B. 11^{12}

   C. 11^8

   D. 11^4

5. Which shows $4^6 \div 4^5$ in standard form?

   A. 0

   B. 1

   C. 4

   D. 16

6. Which shows $2^{-2} \times 2^6$ in exponential form?

   A. 2^4

   B. 2^{-4}

   C. 2^{-8}

   D. 2^{-12}

7. Which shows $(2^2)^{-2}$ in standard form?

   A. 0

   B. $\frac{1}{16}$

   C. $\frac{1}{8}$

   D. 1

8. Which shows $6^{-1} \div 6^{-4}$ in exponential form?

   A. 6^{-5}

   B. 6^{-3}

   C. 6^1

   D. 6^3

9. Look at the expression below.

$$a^6 \div a^4$$

   A. Simplify the expression. Show your work.

   B. Which property of powers did you use to answer Part A?

_____

_____

Square Roots and Cube Roots

Common Core State Standard:
8.EE.2

Getting the Idea

Squaring a number means raising it to the power of 2. For example, 7^2 is equivalent to 7×7, or 49. So, we say that 49 is a perfect square.

The opposite, or inverse, of squaring a number is taking its square root. We use the **radical** symbol ($\sqrt{}$) to represent square roots. To find the square root of a perfect square, think about what number, when multiplied by itself, will result in that perfect square.

Example 1

Solve for y.

$y^2 = 196$

Strategy   **Determine what number, multiplied by itself, results in 196.**

Step 1   Take the square root of both sides of the equation.

The opposite of squaring a number is taking its square root.

$\sqrt{y^2} = \sqrt{196}$

$y = \sqrt{196}$

Step 2   Try squaring numbers until you find one that results in 196.

$12^2 = 12 \times 12 = 144$ Too low

$13^2 = 13 \times 13 = 169$ Too low

$14^2 = 14 \times 14 = 196$ ✓

Step 3   Solve for y.

$14^2 = 196$, so $\sqrt{196} = 14$.

$y = \sqrt{196} = 14$

Solution   **Since $\sqrt{196} = 14$, $y = 14$.**

Cubing a number means raising it to the power of 3. For example, 2^3 is equivalent to $2 \times 2 \times 2$, or 8. So, we say that 8 is a **perfect cube**.

The opposite, or inverse, of cubing a number is taking its **cube root**. We use the symbol $\sqrt[3]{}$ to represent cube roots. To find the cube root of a perfect cube, think about what number, when cubed, will result in that perfect cube.

Example 2

Solve for r.

$$r^3 = 125$$

Strategy	**Determine what number, when cubed, results in 125.**
Step 1	Take the cube root of both sides of the equation.

The opposite of cubing a number is taking its cube root.

$$\sqrt[3]{r^3} = \sqrt[3]{125}$$
$$r = \sqrt[3]{125}$$

Step 2	Try cubing numbers until you find one that gives a result of 125.

You already know that $2^3 = 8$. Start with 3.

$3^3 = 3 \times 3 \times 3 = 27$ Too low
$4^3 = 4 \times 4 \times 4 = 64$ Too low
$5^3 = 5 \times 5 \times 5 = 125$ ✓

Step 3	Solve for r.

$5^3 = 125$, so $\sqrt[3]{125} = 5$.
$$r = \sqrt[3]{125}$$
$$r = 5$$

Solution   Since $\sqrt[3]{125} = 5$, $r = 5$.

The number under a radical sign is called the **radicand**. If you do not have a calculator handy, you may need to estimate the value of a square root or a cube root.

To estimate a square root, find the two perfect squares between which the radicand lies. Take the square root of each to find the range of your estimate.

To estimate a cube root, find the two perfect cubes between which the radicand lies. Then take the cube root of each to find the range of your estimate.

Example 3

Between which two consecutive integers is $\sqrt[3]{500}$?

Strategy	**Find the two perfect cubes between which 500 lies. Then take the cube root of each to make your estimate.**

Step 1 Try cubing consecutive positive integers.

You know from Example 2 that $5^3 = 125$. Start with 6.

$6^3 = 6 \times 6 \times 6 = 216$

$7^3 = 7 \times 7 \times 7 = 343$

$8^3 = 8 \times 8 \times 8 = 512$

The radicand, 500, is between the perfect cubes 343 and 512.

Step 2 Estimate the value of $\sqrt[3]{500}$.

$\sqrt[3]{343} < \sqrt[3]{500} < \sqrt[3]{512}$

$7 < \sqrt[3]{500} < 8$

Solution $\sqrt[3]{500}$ **has a value between 7 and 8.**

Roots can help you solve measurement problems, such as problems involving the area of a square or the volume of a cube.

Coached Example

The area of the square garden on the right is 121 square yards.

What is the length, s, of each side of the garden?

The formula for finding the area, A, of a square is $A = s^2$, where s is the length of a side.

The area of the garden above is _____ square yards.

To find the length of one side, take the _____ root of that area.

On the lines below, try squaring numbers until you find one that gives the result of _____. That is the value of s.

_____

_____

The length of each side, s, of the garden is _____ yards.

Lesson Practice

Choose the correct answer.

1. What is the value of $\sqrt{100}$?

   A. 4

   B. 10

   C. 25

   D. 50

2. What is the value of $\sqrt[3]{27}$?

   A. 3

   B. 5

   C. 9

   D. 13.5

3. Solve for y.

   $$y^3 = 216$$

   A. $y = 4$

   B. $y = 6$

   C. $y = 7$

   D. $y = 15$

4. Between which two consecutive integers is $\sqrt[3]{11}$?

   A. 0 and 1

   B. 1 and 2

   C. 2 and 3

   D. 4 and 5

5. Solve for x.

   $$x^2 = 256$$

   A. $x = 6$

   B. $x = 15$

   C. $x = 16$

   D. $x = 128$

6. Between which two consecutive integers is $\sqrt[3]{200}$?

   A. 66 and 67

   B. 20 and 21

   C. 6 and 7

   D. 5 and 6

7. Which statement below is true?

   A. $\sqrt{1} = \sqrt[3]{1}$

   B. $\sqrt{2} = \sqrt[3]{3}$

   C. $\sqrt{4} = \sqrt[3]{9}$

   D. $\sqrt{4} = \sqrt[3]{27}$

8. Which statement below is true?

   A. $\sqrt{4} = \sqrt[3]{4}$

   B. $\sqrt{4} = \sqrt[3]{27}$

   C. $\sqrt{16} = \sqrt[3]{27}$

   D. $\sqrt{16} = \sqrt[3]{64}$

9. The wooden block shown below is a cube. It has a volume of 512 cubic entimeters.

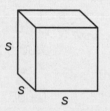

A. What is the length of one side, s? (Hint: the formula for the volume, V, of a cube is $V = s^3$.) Show your work.

_____

B. Indira wants to paint the front face of the block. What is the area of one of the faces? Show your work.

_____

Scientific Notation

Common Core State Standard:
8.EE.4

Getting the Idea

Scientific notation is a way to abbreviate very large or very small numbers using powers of 10. A number written in scientific notation consists of two factors. The first factor is a number greater than or equal to 1, but less than 10. The second factor is a power of 10.

Here are some guidelines and examples of numbers written in scientific notation.

	Standard Form	Scientific Notation
Numbers $\geq$ 10	8,000,000	8×10^6
Numbers $\geq$ 1 and $<$ 10	3	3×10^0
Numbers $>$ 0 and $<$ 1	0.0007	7×10^{-4}

Recall that a number raised to the power of 0 equals 1, so multiplying by 10^0 is the same as multiplying by 1.

Example 1

Jupiter's minimum distance from the sun is about 460,100,000 miles. What is that number written in scientific notation?

Strategy   **Use the definition of scientific notation to find the two factors.**

Step 1   Write the first factor, which must be greater than or equal to 1 and less than 10.

Put the decimal point after the first non-zero digit, starting at the left.

Drop all zeros after the last non-zero digit.

4.601~~00000~~

The first factor is 4.601.

Step 2   Find the exponent for the power of 10.

Count the number of places that the decimal point was moved.

4.60100000

The decimal point was moved 8 places to the left.

Since the original number is greater than 10, the exponent will be positive.

8 is the exponent for the power of 10.

Step 3 Write the second factor.

The exponent is positive 8. The second factor is 10^8.

Step 4 Write the number in scientific notation.

4.601×10^8

Solution **Jupiter's minimum distance from the sun is about 4.601×10^8 miles.**

Example 2

Mr. Kendall measured a specimen that was 0.00000045 millimeter long.
What is the specimen's length written in scientific notation?

Strategy **Use the definition of scientific notation to find the two factors.**

Step 1 Write the first factor, which must be greater than or equal to 1 and less than 10.

Put the decimal point after the first non-zero digit, starting from the left.

Drop the zeros that precede that digit.

0.00000045

The first factor is 4.5.

Step 2 Find the exponent for the power of 10.

Count the number of places that the decimal point was moved.

0.00000045

The decimal point was moved 7 places to the right.

Since the original number is less than 1, the exponent will be negative.

-7 is the exponent for the power of 10.

Step 3 Write the second factor.

The exponent is negative 7. The second factor is 10^{-7}.

Step 4 Write the number in scientific notation.

4.5×10^{-7}

Solution **The specimen was 4.5×10^{-7} millimeter long.**

When converting from scientific notation to standard form, move the decimal point to the right for a positive power of 10 and to the left for a negative power of 10.

Example 3

What is 3.5×10^{-6} written in standard form?

Strategy   **Look at the exponent of the second factor to move the decimal point.**

Step 1   Look at the exponent of the second factor.

The exponent is negative, so the decimal point will move to the left.

The exponent is -6, so move the decimal point 6 places to the left.

Step 2   Move the decimal point in 3.5 six places to the left.

Add zeros as needed.

0.0000035

Step 3   Use a scientific calculator to check your solution.

To find 3.5×10^{-6}:

Type 3.5.   3.5

Press the multiplication sign key. $\boxed{\times}$

Type 10.   10

Press the exponent key. $\boxed{\wedge}$

Type 6.   6

Press the positive/negative key to change the sign on the 6. $\boxed{+/-}$

Press the equal sign key. $\boxed{=}$
The screen should show 0.0000035.
The solution is correct.

Solution $3.5 \times 10^{-6} = 0.0000035$

To multiply numbers in scientific notation, first multiply the decimal factors and then multiply the power-of-10 factors. Remember to use the properties of powers when you multiply the power-of-10 factors. In the example below, *a* and *b* are the decimal factors.

$$(a \times 10^{m})(b \times 10^{n}) = ab \times 10^{m+n}$$

To divide numbers in scientific notation, first divide the decimal factors. Then divide the power-of-10 factors, using the properties of powers. In the example below, a and b are the decimal factors and $b \neq 0$.

$$\frac{(a \times 10^m)}{(b \times 10^n)} = \frac{a}{b} \times 10^{m-n}$$

When you multiply or divide numbers in scientific notation, your product or quotient may not be in scientific notation because the decimal factor is not greater than or equal to 1 and less than 10. To fix this, write the decimal factor in scientific notation and use the properties of powers to simplify the expression.

Example 4

Find the product in scientific notation.

$$(1.5 \times 10^3)(7.8 \times 10^{-7})$$

Strategy   **Multiply the decimal-number factors. Then multiply the power-of-10 factors.**

Step 1   Use the commutative and associative properties to regroup the factors.

$$(1.5 \times 10^3)(7.8 \times 10^{-7}) = (1.5 \times 7.8)(10^3 \times 10^{-7})$$

Step 2   Multiply the decimal factors.

$$1.5 \times 7.8 = 11.7$$

Step 3   Multiply the power-of-10 factors.

$$10^3 \times 10^{-7} = 10^{3 + (-7)} = 10^{-4}$$

Step 4   Write the product using the products from Steps 2 and 3.

$$(1.5 \times 10^3)(7.8 \times 10^{-7}) = 11.7 \times 10^{-4}$$

Step 5   Write 11.7×10^{-4} in scientific notation.

First, write 11.7 in scientific notation.

Move the decimal one place to the left, then multiply the result by 10^1.

11.7 written in scientific notation is 1.17×10^1.

Now, substitute this into the expression and simplify.

$$11.7 \times 10^{-4} = 1.17 \times 10^1 \times 10^{-4}$$

$1.17 \times 10^1 \times 10^{-4} = 1.17 \times (10^1 \times 10^{-4})$ Use the associative property of multiplication.

$1.17 \times (10^1 \times 10^{-4}) = 1.17 \times 10^{1 + (-4)}$ Use the product of powers property.

$$1.17 \times 10^{1 + (-4)} = 1.17 \times 10^{-3}$$

Solution $(1.5 \times 10^3)(7.8 \times 10^{-7}) = 1.17 \times 10^{-3}$

Example 5

What is $4.2 \times (2.5 \times 10^{-6})$ written in standard form?

Strategy   **Use the associative property to regroup the factors. Then write the product in standard form.**

Step 1   Use the associative property to regroup the factors.

$$4.2 \times (2.5 \times 10^{-6}) = (4.2 \times 2.5) \times 10^{-6}$$

Step 2   Multiply the decimal factors.

$$4.2 \times 2.5 = 10.5$$

Step 3   Rewrite the expression using the result from Step 2 and the power-of-10 factor.

$$(4.2 \times 2.5) \times 10^{-6} = 10.5 \times 10^{-6}$$

Step 4   Write the product in standard form.

Look at the power-of-10 factor.

The negative exponent means you move the decimal point to the left.

So, −6 means you move the decimal point in 10.5 six places to the left.

0.0000105

Solution $4.2 \times (2.5 \times 10^{-6}) = 0.0000105$

Example 6

Find the quotient in scientific notation.

$$\frac{8.82 \times 10^5}{3.6 \times 10^3}$$

Strategy   **Divide the decimal-number factors and divide the power-of-10 factors.**

Step 1   Rewrite the expression.

$$\frac{8.82 \times 10^5}{3.6 \times 10^3} = \frac{8.82}{3.6} \times \frac{10^5}{10^3}$$

Step 2   Divide the decimal-number factors.

$$\frac{8.82}{3.6} = 2.45$$

Step 3   Divide the power-of-10 factors.

$$\frac{10^5}{10^3} = 10^{5-3} = 10^2$$

Step 4   Write the result using the quotients from Steps 2 and 3.

$$2.45 \times 10^2$$

Step 5	Check that the product is written in scientific notation.

The first factor, 2.45, is greater than or equal to 1 and less than 10.

The second factor, 10^2, is a power of 10.

The product is written in scientific notation.

Solution $\dfrac{8.82 \times 10^5}{3.6 \times 10^3} = 2.45 \times 10^2$

Coached Example

In 2008, the Hartsfield-Jackson Atlanta International Airport ranked as the world's busiest airport. In that year, approximately 9.0×10^7 passengers passed through this airport. What is that number written in standard form?

Since the exponent is positive, this is a number greater than _____.

The exponent of the second factor is _____.

The exponent tells you to move the decimal point in 9.0 _____ places to the _____.

The number 9.0×10^7 in standard form is _____.

About _____ passengers passed through the Hartsfield-Jackson Atlanta International Airport in 2008.

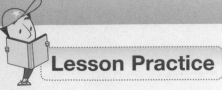

Lesson Practice

Choose the correct answer.

1. What is 0.000058 written in scientific notation?

   A. 5.8×10^{-6}

   B. 5.8×10^{-5}

   C. 5.8×10^{5}

   D. 5.8×10^{6}

2. The length of the Amazon River in South America is 6,400 kilometers. What is this length written in scientific notation?

   A. 6.4×10^{2} km

   B. 6.4×10^{3} km

   C. 6.4×10^{4} km

   D. 6.4×10^{5} km

3. What is 6.92×10^{-3} written in standard form?

   A. 0.000692

   B. 0.00692

   C. 0.0692

   D. 0.692

4. The area of Australia is approximately 7,700,000 square kilometers. What is this area written in scientific notation?

   A. 7.7×10^{-6} sq km

   B. 7.7×10^{-5} sq km

   C. 7.7×10^{5} sq km

   D. 7.7×10^{6} sq km

5. What is 4.01×10^{0} written in standard form?

   A. 0.401

   B. 4.001

   C. 4.01

   D. 40.1

6. A virus is viewed under a microscope. Its diameter is 0.0000002 meter. How would this length be expressed in scientific notation?

   A. 2×10^{-7} meter

   B. 2×10^{-6} meter

   C. 2×10^{6} meters

   D. 2×10^{7} meters

7. Find the product.

$$(1.9 \times 10^3)(4.5 \times 10^2)$$

A. 8.55×10^1

B. 8.55×10^3

C. 8.55×10^5

D. 8.55×10^6

8. Find the quotient.

$$\frac{2.89 \times 10^2}{3.4 \times 10^{-2}}$$

A. 0.85×10^0

B. 0.85×10^4

C. 8.5×10^3

D. 8.5×10^5

9. Mohammed copied this problem into his notebook.

$$(3.4 \times 10^5)(3.8 \times 10^{-9})$$

A. Use the associative and commutative properties to rearrange the factors.

_____

B. Find the product. Write the product in standard form.

Solve Problems Using Scientific Notation

Common Core State Standards:
8.EE.3, 8.EE.4

Getting the Idea

Sometimes, you may need to multiply or divide numbers written in scientific notation in order to solve real-world problems.

Example 1

A rectangular section of wilderness will be set aside as a new wildlife refuge. Its dimensions are 5×10^5 meters by 4×10^4 meters. Find the area of the land in square meters. Then convert the area into square kilometers using the conversion below.

1 square kilometer (km^2) = 1×10^6 square meters (m^2)

Which unit is a better choice for measuring the area of the wildlife refuge, and why?

Strategy **Multiply the dimensions. Convert the area into square kilometers. Compare the two units.**

Step 1 Multiply the dimensions to find the area, A.

A (in m^2) = $(5 \times 10^5 \text{ m})(4 \times 10^4 \text{ m})$

Step 2 Multiply the first factors and then multiply the power-of-10 factors.

$5 \times 4 = 20$

$10^5 \times 10^4 = 10^{5+4} = 10^9$

So, A (in m^2) = $(5 \times 10^5 \text{ m})(4 \times 10^4 \text{ m}) = 20 \times 10^9$.

Step 3 Rewrite the number in scientific notation.

$20 \times 10^9 = 2 \times 10^1 \times 10^9 = 2 \times 10^{1+9} = 2 \times 10^{10}$

Step 4 Convert the area into square kilometers.

To convert a smaller unit (square meters) to a larger unit (square kilometers), divide:

A (in km^2) = $\dfrac{2 \times 10^{10}}{1 \times 10^6}$

Divide the first factors and then divide the power-of-10 factors.

$\dfrac{2}{1} = 2$

$\dfrac{10^{10}}{10^6} = 10^{10-6} = 10^4$

So, A (in km^2) = 2×10^4.

Step 5   Which is the better unit to use?

$$2 \times 10^{10} \text{ m}^2 = 20{,}000{,}000{,}000 \text{ m}^2$$

$$2 \times 10^4 \text{ km}^2 = 20{,}000 \text{ km}^2$$

20,000 is a more reasonable number to work with in standard form. Also, square kilometers are larger units than square meters. Since the area is large, it is better to use the larger unit.

Solution   **The area of the refuge is 2×10^{10} square meters or 2×10^4 square kilometers. Square kilometers is a better unit to use because the area is very large.**

Sometimes, you may use technology, such as a calculator, to generate a number. If the result is a number that is very large or very small, many calculators will automatically give the number in scientific notation.

Example 2

One cubic millimeter of Ms. Murphy's blood contains about 5,000,000 red blood cells. There are about 4,900,000 cubic millimeters of blood in her entire body. Use a calculator to determine approximately how many red blood cells Ms. Murphy has in total. Interpret the number your calculator gives as the final answer.

Strategy   **Use a calculator to determine the answer. Interpret the result.**

Step 1   How can you find the total number of red blood cells?

Multiply the number of red blood cells in one cubic millimeter of blood (5,000,000) by the total number of cubic millimeters of blood in the body (4,900,000).

Step 2   Use a calculator to determine the answer.

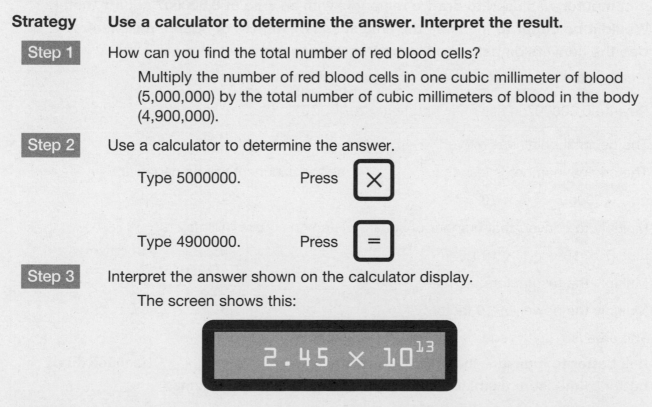

Type 5000000.   Press $\boxed{\times}$

Type 4900000.   Press $\boxed{=}$

Step 3   Interpret the answer shown on the calculator display.

The screen shows this:

$$2.45 \times 10^{13}$$

Solution   **Ms. Murphy has a total of about 2.45×10^{13} red blood cells in her body.**

Example 3

California, the most populous state, has approximately 4×10^7 people living in it.

The population of the entire United States is approximately 3×10^8 people. About how many times greater is the population of the United States than the population of California?

Strategy Decide if you should multiply or divide. Then solve the problem.

Step 1 Decide on which operation to use.

To find how many times greater, divide 3×10^8 by 4×10^7.

Step 2 Divide the first factors and then divide the power-of-10 factors.

$$\frac{3}{4} = 0.75 \qquad\qquad \frac{10^8}{10^7} = 10^{8-7} = 10^1 = 10$$

So, $\frac{3 \times 10^8}{4 \times 10^7} = 0.75 \times 10 = 7.5$

Solution The population of the United States is $7\frac{1}{2}$ times the population of California.

Coached Example

A computer was used to draw a rectangle with an area of 0.000007 square meter. Would it be better to measure the area in square meters or square millimeters? Use the conversion below to help determine your answer.

1 square meter (m^2) = 1×10^6 square millimeters (mm^2)

Rewrite 0.000007 in scientific notation.   0.000007

The decimal point was moved _____ places to the right.

The original number is less than _____, so the exponent will be negative.

$0.000007 = 7 \times 10^{-}$ ___

Multiply to convert that number of square meters to square millimeters:

$(7 \times 10^{-} \text{___})(1 \times 10^6)$

Multiply the first factors: $7 \times 1 = $ _____

Multiply the power-of-10 factors: _____

The area is _____ square millimeters.

It is better to measure the area in square _____ because it is better to measure a small area using a _____ unit.

Lesson Practice

Choose the correct answer.

1. One microgram is equal to 1×10^{-6} gram. If the mass of a substance is 8×10^{9} micrograms, what is its mass in grams?

   A. 1.25×10^{-15} gram

   B. 1.25×10^{-3} gram

   C. 8×10^{3} grams

   D. 8×10^{15} grams

2. A rectangular section of land made up of wheat farms has a length of 5×10^{4} meters and a width of 6×10^{3} meters. What is the area of the land in square meters?

   A. 3×10^{6} square meters

   B. 3×10^{7} square meters

   C. 3×10^{8} square meters

   D. 3×10^{12} square meters

3. A microscope is set so it makes an object appear 4×10^{2} times larger than its actual size. A virus has a diameter of 2×10^{-7} meter. How large will the diameter of the virus appear when it is viewed under the microscope?

   A. 8×10^{-14} meter

   B. 8×10^{-5} meter

   C. 8×10^{-4} meter

   D. 8×10^{5} meters

4. Neptune is approximately 5×10^{4} kilometers in diameter. Mars is approximately 7×10^{3} kilometers in diameter. Which is an accurate comparison of the diameters of these two planets?

   A. The diameter of Neptune is more than 7 times greater than the diameter of Mars.

   B. The diameter of Mars is more than 7 times greater than the diameter of Neptune.

   C. The diameter of Neptune is about 1.4 times greater than the diameter of Mars.

   D. The diameter of Mars is about 1.4 times greater than the diameter of Neptune.

5. A box contains 5×10^{3} paper clips. The mass of each paper clip in the box is 8×10^{-4} kilogram. What is the combined mass of the paper clips in the box?

   A. 4 kilograms

   B. 40 kilograms

   C. 4×10^{7} kilograms

   D. 4×10^{-7} kilogram

6. The head of a pin has a diameter of 1×10^{-4} meter. A bacterium has a diameter of 5×10^{-7} meter. How many bacteria that size would fit across the diameter of the pinhead?

   A. 2   **C.** 200

   B. 20   **D.** 2×10^{11}

7. The population of Canada is approximately 3×10^7. The population of Mexico is approximately 1×10^8. Which statement accurately compares the populations of Canada and Mexico?

A. The population of Canada is more than 30 times greater than the population of Mexico.

B. The population of Mexico is more than 30 times greater than the population of Canada.

C. The population of Canada is more than 3 times greater than the population of Mexico.

D. The population of Mexico is more than 3 times greater than the population of Canada.

8. One nanometer is equivalent to 1×10^{-9} meters. Which is equivalent to 0.3 nanometers?

A. -3×10^{10} meters

B. -3^{10} meters

C. 3×10^{-9} meter

D. 3×10^{-10} meter

9. A rectangular yard has a length of 0.006 kilometer and a width of 5×10^{-3} kilometer.

A. Use scientific notation to express the area of the yard in square kilometers, showing each step in the process.

B. Convert the area into square meters using the conversion below.

1 square kilometer (km^2) = 1×10^6 square meters (m^2)

Give your answer in standard form. Which unit is a better choice for measuring the area of the yard, and why?

_____

_____

_____

_____

Linear Equations in One Variable

Common Core State Standards:
8.EE.7.a, 8.EE.7.b

Getting the Idea

An **equation** is a mathematical sentence that uses an equal (=) sign to show that two quantities are equal in value. A **variable** is a symbol or letter that is used to represent one or more numbers.

A **linear equation** has one or more variables raised to the first power. You can use **inverse operations** and the properties of equality to solve a linear equation that has one variable.

Properties of Equality	
Addition Property of Equality	**Multiplication Property of Equality**
If you add the same number to both sides of an equation, the equation continues to be true. If $a = c$, then $a + b = c + b$.	If you multiply both sides of an equation by the same number, the equation continues to be true. If $a = c$, then $ab = cb$.
Subtraction Property of Equality	**Division Property of Equality**
If you subtract the same number from both sides of an equation, the equation continues to be true. If $a = c$, then $a - b = c - b$.	If you divide both sides of an equation by the same nonzero number, the equation continues to be true. If $a = c$ and $b \neq 0$, then $\frac{a}{b} = \frac{c}{b}$.

The key is to remember that whatever you do to one side of the equation, you must also do to the other side. That way, you can **isolate the variable** while still keeping the equation true.

Example 1

Find the value of x in this equation.

$$-\frac{x}{6} + 9 = -1$$

Strategy   **Use inverse operations and one or more properties of equality.**

Step 1   Get $-\frac{x}{6}$ by itself on one side of the equation.

Since 9 is added to $-\frac{x}{6}$, subtract 9 from both sides.

$$-\frac{x}{6} + 9 = -1$$

$$-\frac{x}{6} + 9 - 9 = -1 - 9$$

$$-\frac{x}{6} = -10$$

The equation is still true because of the subtraction property of equality.

Step 2 Get x by itself.

Since x is divided by -6, multiply both sides by -6.

$$-\frac{x}{6} = -10$$

$$-\frac{x}{6} \cdot -\frac{6}{1} = -10 \cdot -6$$

$$x = 60$$

The equation is still true because of the multiplication property of equality.

Step 3 Check your answer.

Substitute 60 for x in the original equation.

$$-\frac{x}{6} + 9 = -1$$

$$-\frac{60}{6} + 9 \stackrel{?}{=} -1$$

$$-10 + 9 \stackrel{?}{=} -1$$

$$-1 = -1 \checkmark$$

Note: You can check your answer for any linear equation you solve.

Solution **The value of x is 60.**

To undo multiplication by a fraction, multiply by the reciprocal of that fraction.

Flip the fraction to find its reciprocal. For example, the reciprocal of $\frac{4}{5}$ is $\frac{5}{4}$.

Example 2

What is the value of c in this equation?

$$\frac{2}{3}c - \frac{3}{5} = \frac{7}{10}$$

Strategy **Use inverse operations and one or more properties of equality.**

Step 1 Get $\frac{2}{3}c$ by itself on one side of the equation.

Since $\frac{3}{5}$ is subtracted from $\frac{2}{3}c$, add $\frac{3}{5}$ to both sides.

$$\frac{2}{3}c - \frac{3}{5} = \frac{7}{10}$$

$$\frac{2}{3}c - \frac{3}{5} + \frac{3}{5} = \frac{7}{10} + \frac{3}{5}$$

$$\frac{2}{3}c = \frac{7}{10} + \frac{3}{5} \qquad \text{Give } \frac{7}{10} \text{ and } \frac{3}{5} \text{ the same denominator.}$$

$$\frac{2}{3}c = \frac{7}{10} + \frac{6}{10} \qquad \frac{3}{5} = \frac{3 \times 2}{5 \times 2} = \frac{6}{10}$$

$$\frac{2}{3}c = \frac{13}{10}$$

Step 2   Get c by itself.

Since $\frac{2}{3}$ is multiplied by c, divide both sides by $\frac{2}{3}$ or multiply by the reciprocal, $\frac{3}{2}$.

$$\frac{2}{3}c = \frac{13}{10}$$

$$\frac{2}{3}c \cdot \frac{3}{2} = \frac{13}{10} \cdot \frac{3}{2}$$

$$1c = \frac{39}{20}$$

$$c = 1\frac{19}{20}$$

Solution   The value of c is $\frac{39}{20}$ or $1\frac{19}{20}$.

You may also need to combine **like terms** if the same variable is on both sides of the equation. Like terms are terms that contain the same variables raised to the same power.

Example 3

What is the value of z in this equation?

$$0.8z + 3.74 = z + 1.5$$

Strategy   **Combine like terms. Then solve.**

Step 1   Combine like terms so there is only one variable term.

Subtract $0.8z$ from both sides.

$$0.8z + 3.74 = z + 1.5$$

$0.8z - 0.8z + 3.74 = z - 0.8z + 1.5$ Remember: $z - 0.8z = 1.0z - 0.8z$

$$3.74 = 0.2z + 1.5$$

Now, $0.2z$ is the only variable term in the equation.

Step 2   Get $0.2z$ by itself on one side of the equation.

Subtract 1.5 from both sides.

$$3.74 = 0.2z + 1.5$$

$$3.74 - 1.5 = 0.2z + 1.5 - 1.5$$

$$3.74 - 1.50 = 0.2z$$

$$2.24 = 0.2z$$

Step 3   Solve for z.

Divide both sides by 0.2.

$$2.24 = 0.2z$$

$$\frac{2.24}{0.2} = \frac{0.2z}{0.2}$$

$$11.2 = z$$

Solution   The value of z is 11.2.

To solve some equations, you may need to use the distributive property.

> ## Distributive Property
>
> When a factor is multiplied by the sum of two numbers, you can multiply each of the two numbers by the factor and then add the products.
>
> $$a(b + c) = ab + ac$$
>
> When a factor is multiplied by the difference of two numbers, you can multiply each of the two numbers by the factor and then subtract the products.
>
> $$a(b - c) = ab - ac$$

Example 4

What is the value of y in this equation?

$$4y - 1 = 2(y - 2)$$

Strategy   **Use the distributive property. Then combine like terms.**

Step 1   Apply the distributive property to evaluate the expression on the right side of the equal sign.

Distribute the 2 over $(y - 2)$.

$$4y - 1 = 2(y - 2)$$
$$4y - 1 = (2 \cdot y) - (2 \cdot 2)$$
$$4y - 1 = 2y - 4$$

Step 2   Combine like terms so there is only one variable term in the equation.

Subtract $2y$ from both sides.

$$4y - 1 = 2y - 4$$
$$4y - 1 - 2y = 2y - 4 - 2y$$
$$2y - 1 = -4$$

Now, $2y$ is the only variable term in the equation.

Step 3   Get $2y$ by itself on one side of the equation.

Add 1 to both sides.

$$2y - 1 = -4$$
$$2y - 1 + 1 = -4 + 1$$
$$2y = -3$$

Step 4 Get y by itself on one side of the equation.

Divide both sides by 2.

$$2y = -3$$
$$\frac{2y}{2} = -\frac{3}{2}$$
$$y = -\frac{3}{2}$$
$$y = -1\frac{1}{2}$$

Solution The value of y is $-\frac{3}{2}$ or $-1\frac{1}{2}$.

In Examples 1 through 4, all the equations had one solution. However, linear equations may also have no solution or infinitely many solutions.

One Solution	No Solution	Infinitely Many Solutions
Only one number, 1, makes the equation below true. Example: $2x = x + 1$ $x = 1$	No number makes the equation below true. Example: $x + 1 = x + 2$ $1 \neq 2$ → never true	Any number makes the equation below true. Example: $x + 0 = x$ $x = x$ → always true

Example 5

Does this equation have one solution, no solutions, or infinitely many solutions?

$$10q - 15 = 5(2q + 4)$$

Strategy **Use the distributive property. Then combine like terms.**
Decide how many solutions the equation has.

Step 1 Apply the distributive property to evaluate the expression on the right side of the equal sign.

Distribute 5 over $(2q + 4)$.

$$10q - 15 = 5(2q + 4)$$
$$10q - 15 = (5 \cdot 2q) + (5 \cdot 4)$$
$$10q - 15 = 10q + 20$$

Step 2 Combine like terms.

Subtract $10q$ from both sides.

$$10q - 15 = 10q + 20$$
$$10q - 10q - 15 = 10q - 10q + 20$$
$$-15 = 20 \quad \text{→ never true}$$

Step 3 Determine the solution of the equation.

Since $-15 \neq 20$, no value of q makes the equation true.

The equation has no solutions.

Solution   **The equation has no solutions.**

Coached Example

Does the equation below have one solution, no solutions, or infinitely many solutions?

$$n + 2 = \tfrac{1}{3}(3n + 6)$$

Apply the distributive property.

Distribute $\tfrac{1}{3}$ over $(3n + 6)$.

$$n + 2 = \tfrac{1}{3}(3n + 6)$$

$$n + 2 = \left(\tfrac{1}{3} \cdot \tfrac{3n}{1}\right) + \left(\tfrac{1}{3} \cdot \underline{\quad}\right)$$

$$n + 2 = \underline{\quad} + \underline{\quad}$$

Subtract 2 from both sides.

$$n + 2 - 2 = \underline{\qquad\qquad} - 2$$

$$n = \underline{\quad}$$

Is the equation above always true, never true, or sometimes true?

The equation is _____ true, so any value of n makes the equation true.

Does the equation have one solution, no solutions, or infinitely many solutions?

Since any value of n makes the equation true, the equation has _____ solution(s).

Lesson Practice

Choose the correct answer.

1. What is the solution for $72x + 7 = 223$?

   A. $x = 6$

   B. $x = 4$

   C. $x = 3$

   D. $x = 2$

2. Which best describes the solution for $\frac{g}{2} - 6 = 4$?

   A. $g = 20$

   B. $g = 5$

   C. no solution

   D. infinitely many solutions

3. What value of u makes the equation true?

   $$u - 9 = -7u + 7$$

   A. $u = 2$

   B. $u = 2\frac{2}{3}$

   C. $u = 16$

   D. $u = 32$

4. What value of x makes the equation true?

   $$\frac{3}{4}x + 9 = 3$$

   A. $x = -8$

   B. $x = -\frac{1}{2}$

   C. $x = 1$

   D. $x = 16$

5. What value of t makes this equation true?

   $$6t - 8 = 2(2t + 1)$$

   A. $t = -3$

   B. $t = 1$

   C. $t = 2$

   D. $t = 5$

6. Which best describes the solution for the equation below?

   $$3\left(4k - \frac{1}{3}\right) = 12k - 1$$

   A. $k = -\frac{1}{5}$

   B. $k = -\frac{7}{15}$

   C. no solution

   D. infinitely many solutions

7. Which best describes the solution for the equation below?

$$0.5(2x + 8) = x - 4$$

A. $x = -4$

B. $x = 0$

C. no solution

D. infinitely many solutions

8. What value of r makes the equation true?

$$\tfrac{1}{4}(4r - 1) = 2r + \tfrac{1}{8}$$

A. $r = -\tfrac{3}{8}$

B. $r = -\tfrac{1}{6}$

C. $r = \tfrac{1}{4}$

D. $r = \tfrac{1}{2}$

9. Marcy made an error when solving the equation below.

$$8m - 20 = 36$$
$$8m - 20 + 20 = 36$$
$$8m = 36$$
$$\frac{8m}{8} = \frac{36}{8}$$
$$m = 4\frac{4}{8}$$
$$m = 4\frac{1}{2}$$

A. Identify Marcy's error and use one or more number properties to explain why it resulted in an incorrect solution.

_____

_____

_____

B. Correctly solve $8m - 20 = 36$ for m. Show your work.

Use One-Variable Linear Equations to Solve Problems

Getting the Idea

Sometimes, you can solve a real-world problem by writing a linear equation to represent it and then solving the equation. Key words can help you translate many, but not all, math problems. Always think about which operation makes sense for a particular problem.

Key Words

Addition	Multiplication	Subtraction	Division
more than plus additional sum	per times of product	less less than fewer than difference	half share equally quotient separate into equal groups
How many altogether? How many in all? What is the total number?		How many more? How many fewer? How many left?	How many in each?

Example 1

Diego has 60 CDs. This is 12 more CDs than Heidi has.
How many CDs does Heidi have?

Strategy   **Use key words to translate the problem into an equation. Then solve.**

Step 1   Identify the key words.

The words "more than" indicate addition.

The word "is" may indicate where to place the equal sign.

Step 2   Translate the words into an equation.

Let h represent the number of CDs Heidi has.

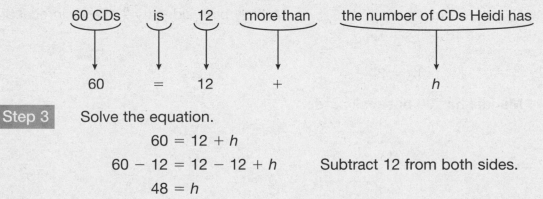

60 CDs	is	12	more than	the number of CDs Heidi has
60	=	12	+	h

Step 3   Solve the equation.

$$60 = 12 + h$$
$$60 - 12 = 12 - 12 + h \qquad \text{Subtract 12 from both sides.}$$
$$48 = h$$

Step 4   Check your answer.

Substitute 48 for *h* into the original equation.

$60 = 12 + h$

$60 \overset{?}{=} 12 + 48$

$60 = 60$ ✓

Solution   **Heidi has 48 CDs.**

Example 2

Simone has six less than $\frac{2}{3}$ of the number of baseball cards that Manuel has.
Simone has 14 baseball cards. How many cards does Manuel have?

Strategy   **Use key words to translate the problem into an equation. Then solve.**

Step 1   Identify the key words.

The words "less than" indicate subtraction.

To show "$\frac{2}{3}$ of", you can multiply by $\frac{2}{3}$.

Step 2   Translate the words into an equation.

Let *m* represent the number of cards that Manuel has.

Simone has six less than $\frac{2}{3}$ as many cards as Manuel.   Simone has 14 cards.

$$\frac{2}{3}m - 6 \qquad = \qquad 14$$

Step 3   Solve the equation.

$\frac{2}{3}m - 6 = 14$

$\frac{2}{3}m - 6 + 6 = 14 + 6$ Add 6 to both sides.

$\frac{2}{3}m = 20$

$\frac{2}{3}m \cdot \frac{3}{2} = \frac{20}{1} \cdot \frac{3}{2}$ Multiply both sides by $\frac{3}{2}$, the reciprocal of $\frac{2}{3}$.

$m = \frac{60}{2}$

$m = 30$

Solution   **Manuel has 30 baseball cards.**

Example 3

A taxi charges $2.50 for each ride plus $1.25 per mile traveled.
If the total charge for one ride was $8.75, how many miles were traveled?

Strategy   **Translate the problem into an equation. Then solve.**

Step 1   Identify any key words or helpful words.

The key word "plus" indicates addition.

The key word "per" indicates multiplication.

Step 2   Translate the words into an equation.

Let m represent the number of miles traveled.

$2.50	plus	$1.25 per mile	total charge of $8.75
2.50	+	1.25 m	=	8.75

Step 3   Solve the equation.

$$2.50 + 1.25m = 8.75 \qquad \text{Subtract 2.50 from each side.}$$
$$1.25m = 6.25 \qquad \text{Divide both sides by 1.25.}$$
$$m = 5$$

Solution   **The taxi traveled 5 miles.**

Coached Example

At a video store, for every DVD bought at the regular price, a customer can buy a second DVD for half the regular price. Nadia buys two DVDs, each of which regularly costs d dollars, and pays $33 in all. What is the regular price of each DVD?

The key word "half" means _____ by 2. Translate the problem into an equation.

One DVD at the regular price of d dollars	Second DVD at half the regular price of d dollars	$33 in all
_____	+	_____	=	_____

Rewrite the equation and solve for d.

The regular price of a DVD at the store is $ _____ .

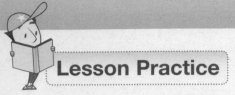

Lesson Practice

Choose the correct answer.

1. Jeremy has 245 stamps. Jonas has 35 fewer stamps than Jeremy has. How many stamps does Jonas have?

    A. 7

    B. 210

    C. 215

    D. 280

2. Harry buys American cheese at a deli. He buys $\frac{1}{4}$ pound of American cheese at a cost of x dollars per pound. If he pays $12.25 for 2.5 pounds of American cheese, what is the cost per pound of the cheese?

    A. $4.90

    B. $6.22

    C. $9.95

    D. $31.12

3. Charlie has a $30 smoothie gift card. Each time he uses the card to buy a smoothie, $2 is deducted from his card. If the balance on his card today is $12, how many smoothies has he purchased with the card?

    A. 3

    B. 6

    C. 9

    D. 18

4. Maneesh and Beth are using leather to make bracelets. Maneesh has y inches of leather. Beth has 7 more inches of leather than Maneesh has. They have a combined total of 77 inches of leather. What is the value of y?

    A. 7 inches

    B. 35 inches

    C. 42 inches

    D. 70 inches

5. Sylvie's age is 5 years less than half Katie's age. If Sylvie is 11 years old, what is Katie's age?

    A. 8 years old

    B. 12 years old

    C. 27 years old

    D. 32 years old

6. The number of birds that Vikram saw on a nature walk is 1 less than $\frac{4}{5}$ the number of birds that Shaya saw. Shaya saw a total of 10 birds. How many birds did Vikram see?

    A. 7

    B. 8

    C. 9

    D. 11

7. Julio bought a stereo that cost $412 and a CD storage case that cost x dollars. He paid the clerk $500 in cash and received $3 in change. Assuming there was no tax, how much did the CD storage case cost?

   A. $415

   B. $409

   C. $88

   D. $85

8. Yolanda has $38 in a bank account. She wants to make two equal deposits so that her account will have a balance of $100. How much money does Yolanda need to deposit each time?

   A. $19

   B. $31

   C. $38

   D. $62

9. For each babysitting job, Ashley charges $2.50 for bus fare plus $8 per hour for each hour she works. She charged $30.50 for her last babysitting job.

   A. Write a linear equation to represent the problem. Be sure to define the variable you choose.

   _____

   _____

   B. How many hours did she babysit for that job? Show every step you used to solve the problem.

Slope

Common Core State Standard:
8.EE.6

Getting the Idea

The graph of a linear equation is a straight line. The steepness of the line is called its **slope**. The slope shows the rate at which two quantities are changing. Specifically, it is the **ratio** of the vertical change to the horizontal change, or $\frac{rise}{run}$.

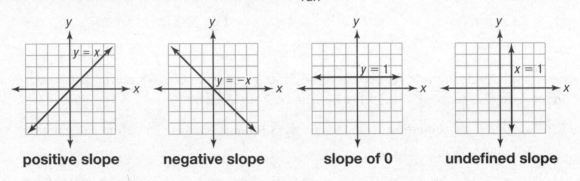

| positive slope | negative slope | slope of 0 | undefined slope |

Look at the lines graphed above.

- A line that slants up from left to right has a positive slope.

  For this graph, as the x-values increase, the y-values also increase.

- A line that slants down from left to right has a negative slope.

  For this graph, as the x-values increase, the y-values decrease.

- A horizontal line has a slope of 0 because there is no vertical change.

- A vertical line has an undefined slope, because there is no horizontal change.

  A fraction with a denominator of 0 is undefined, and $\frac{rise}{run}$ would have a denominator of 0.

Example 1

What is the slope of this line?

Strategy **Find the vertical change and the horizontal change.**

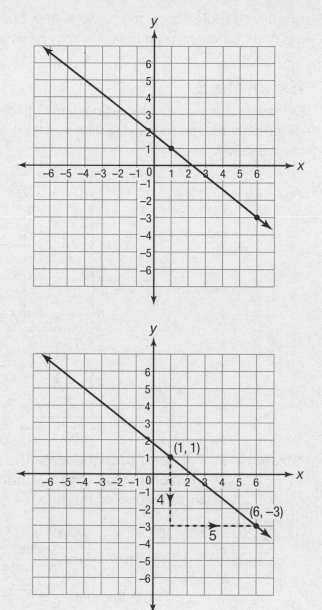

Step 1 Choose two points on the line.

(1, 1) and (6, −3)

Step 2 Identify the vertical change and horizontal change.

From (1, 1) to (6, −3), the line moves 4 units down and 5 units to the right.

Step 3 Write the slope.

The line slants down from left to right, so the slope is negative.

$$\text{slope} = \frac{\text{rise}}{\text{run}} = -\frac{4}{5}$$

Solution The slope is $-\frac{4}{5}$.

Similar triangles have the same shape, but not necessarily the same size. The ratios of the lengths of their corresponding sides are equal.

Example 2

Does the slope of a non-vertical line change depending on which two points you use to determine it? Use the two similar triangles and the line graphed below to help you answer the question.

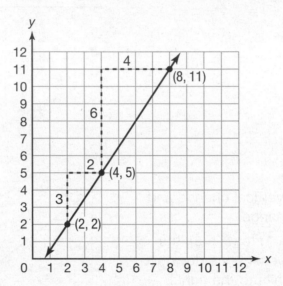

Strategy   For each triangle, write the ratio of the vertical side length to the horizontal side length. Compare those ratios, and compare them to the slope of the line.

Step 1   Find the ratio of the vertical and horizontal side lengths for each triangle.

Smaller triangle: $\dfrac{\text{vertical side length}}{\text{horizontal side length}} = \dfrac{3}{2}$

Larger triangle: $\dfrac{\text{vertical side length}}{\text{horizontal side length}} = \dfrac{6}{4}$

Simplify: $\dfrac{6}{4} = \dfrac{6 \div 2}{4 \div 2} = \dfrac{3}{2}$

Step 2   Compare the ratios.

The ratio of the lengths of the vertical and horizontal sides for both triangles is $\dfrac{3}{2}$.

Step 3 Compare the ratios to the slope.

The slope is the ratio $\frac{\text{rise}}{\text{run}}$ or $\frac{\text{vertical change}}{\text{horizontal change}}$.

If you use the points (2, 2) and (4, 5), slope $= \frac{\text{rise}}{\text{run}} = \frac{3}{2}$.

If you use the points (4, 5) and (8, 11), slope $= \frac{\text{rise}}{\text{run}} = \frac{6}{4} = \frac{3}{2}$.

Step 4 Analyze the slope.

No matter which two points on the line you use, the slope is the same.

This is because the graph of a line changes at a constant rate.

Solution **The slope of a non-vertical line stays the same no matter which two points you use to determine it.**

Example 2 demonstrates that the slope of a line represents a constant **rate of change**. If you know any two points on a line, you can determine its slope using the formula below.

> **Slope**
>
> The slope, m, of a line containing points (x_1, y_1) and (x_2, y_2) is:
>
> $$m = \frac{\text{change in } y}{\text{change in } x} = \frac{y_2 - y_1}{x_2 - x_1}$$

Example 3

What is the slope of a line that passes through $(-3, 6)$ and $(5, 1)$?

Strategy **Use the slope formula.**

Step 1 Identify the points.

Let $(x_1, y_1) = (-3, 6)$.

Let $(x_2, y_2) = (5, 1)$.

Step 2 Substitute the numbers into the slope formula.

$$m = \frac{y_2 - y_1}{x_2 - x_1}$$
$$m = \frac{1 - 6}{5 - (-3)}$$
$$m = \frac{-5}{8}$$
$$m = -\frac{5}{8}$$

Solution The slope is $-\frac{5}{8}$.

The slope of a line represents a constant rate of change. For example, a slope of $\frac{50}{3}$ could represent these rates:

$\dfrac{\$50}{3 \text{ hours}}$ $\qquad\qquad$ $\dfrac{50 \text{ miles}}{3 \text{ gallons}}$ $\qquad\qquad$ $\dfrac{50 \text{ pages}}{3 \text{ minutes}}$

When a graph shows a real-world situation, think about what the slope represents.

Example 4

A locksmith charges a flat fee for each house call plus an hourly rate, as shown by the graph below.

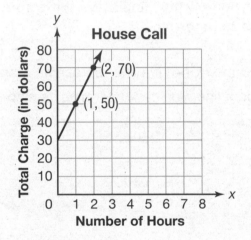

Find the slope of the graph. What does the slope represent in this problem?

Strategy **Find and interpret the slope of the line.**

Step 1 Find the slope.

Let $(x_1, y_1) = (1, 50)$.

Let $(x_2, y_2) = (2, 70)$.

$$m = \frac{y_2 - y_1}{x_2 - x_1} = \frac{70 - 50}{2 - 1} = \frac{20}{1} = 20$$

Step 2 Interpret the slope.

The slope shows the ratio $\dfrac{\text{change in } y}{\text{change in } x}$.

So, the slope compares the total charge, in dollars, to the number of hours worked.

The slope shows a rate of change of $\dfrac{20 \text{ dollars}}{1 \text{ hour}}$, or $20 per hour.

So, the slope represents the hourly rate charged by the locksmith.

Solution **The slope is 20. It shows that the locksmith charges $20 per hour for each job.**

Coached Example

Joanie bought an airplane phone card that charges her a connection fee plus an additional rate for each minute a call lasts. The graph below represents this situation.

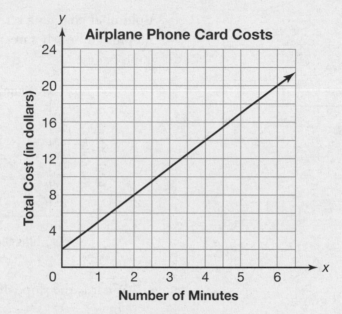

What is the slope of the graph, and what does it represent?

Choose any two points on the graph.

Let $(x_1, y_1) = (2,$ _____ $)$.

Let $(x_2, y_2) = (6,$ _____ $)$.

$m = \dfrac{y_2 - y_1}{x_2 - x_1} = \dfrac{\underline{\quad} - \underline{\quad}}{6 - 2} = \dfrac{\underline{\quad}}{\underline{\quad}} = \underline{\quad}$

The y-axis shows the total cost in _____.

The x-axis shows the time in _____.

So, the slope shows a rate of change of _____ dollars to _____, or _____ per minute.

Does the slope represent the connection fee or the rate per minute for the call? _____

The slope is _____. It shows that Joanie must pay $_____ per minute for the calls she makes.

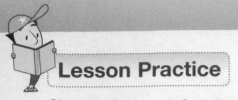

Lesson Practice

Choose the correct answer.

1. What is the slope of this line?

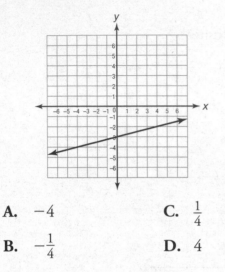

   A. -4 C. $\frac{1}{4}$

   B. $-\frac{1}{4}$ D. 4

2. Which is true of this graph?

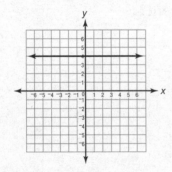

   A. The slope is 0.

   B. The slope is 1.

   C. The slope is 4.

   D. The slope is undefined.

3. What is the slope of a line that passes through $(2, -5)$ and $(6, -2)$?

   A. $-\frac{4}{3}$ C. $\frac{3}{4}$

   B. $-\frac{3}{4}$ D. $\frac{4}{3}$

Use the graph for questions 4 and 5.

A plumber charges a set fee for each house call plus an hourly rate, as shown by the graph below.

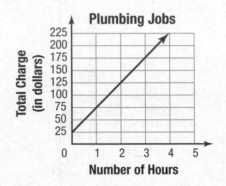

4. What is the slope of the line graphed above?

   A. 100

   B. 75

   C. 50

   D. 25

5. What does the slope of the graph represent?

   A. the cost of materials for each plumbing job

   B. the total charge for any plumbing job

   C. the set fee for any plumbing job

   D. the hourly rate charged by the plumber

6. What is the slope of this line?

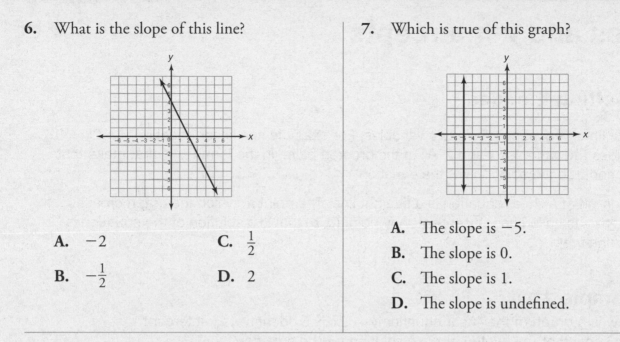

A. -2 **C.** $\frac{1}{2}$

B. $-\frac{1}{2}$ **D.** 2

7. Which is true of this graph?

A. The slope is -5.

B. The slope is 0.

C. The slope is 1.

D. The slope is undefined.

8. Consider the line graphed below.

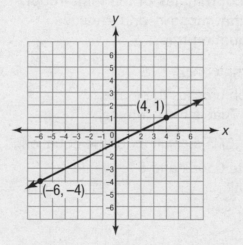

A. Using the two points labeled on the graph above, calculate its slope. Show your work.

B. Does the slope of the line change depending on which two points you use? Calculate the slope using two different points to support your answer.

Slopes and *y*-intercepts

Common Core State Standards:
8.EE.5, 8.EE.6

Getting the Idea

Some linear equations have two variables. For example, the linear equation $y = 2x + 5$ includes the variables x and y. All of the ordered pairs, in the form (x, y), that make that equation true are solutions of the equation.

The graph of a linear equation is a straight line. The point at which the graph crosses the *y*-axis is called its **y-intercept**. Any point $(0, b)$ that is a solution of the equation is the *y*-intercept.

Example 1

Below is a graph of the linear equation $y = 2x + 5$. Identify its *y*-intercept. Then show that the *y*-intercept is a solution for the equation.

Strategy   **Identify the coordinates of the *y*-intercept. Then show that those *x*- and *y*-values make the equation true.**

Step 1   Identify the *y*-intercept.

The graph crosses the *y*-axis at $(0, 5)$. That is the *y*-intercept.

Step 2   Show that $(0, 5)$ is a solution for the equation.

Substitute 0 for *x* and 5 for *y* into the equation.

$$y = 2x + 5$$
$$5 \stackrel{?}{=} 2(0) + 5$$
$$5 \stackrel{?}{=} 0 + 5$$
$$5 = 5 \quad ✓$$

Solution   **The *y*-intercept is $(0, 5)$. Those coordinates are a solution for the equation, as shown in Step 2.**

The equation $y = 2x + 5$ is written in a special form called the **slope-intercept form**.

If a linear equation is in slope-intercept form, you can use the y-intercept and the slope to graph it.

> The slope-intercept form of an equation is:
>
> $y = mx + b$, where m represents the slope and b represents the y-intercept.

Example 2

Graph the equation $y = \frac{2}{3}x + 1$.

Strategy   **Identify the y-intercept and slope. Use the slope to find a second point on the line.**

Step 1   Identify the slope and the y-intercept.

The equation $y = \frac{2}{3}x + 1$ is in slope-intercept form, $y = mx + b$.

$m = \frac{2}{3}$, so the slope is $\frac{2}{3}$.

$b = 1$, so the y-intercept is $(0, 1)$.

Step 2   Use the slope to find a second point.

Plot a point at the y-intercept, $(0, 1)$.

Use the slope to find a second point.

$$\text{slope} = \frac{\text{rise}}{\text{run}} = \frac{2}{3}$$

Start at $(0, 1)$. Since the slope is positive, rise up 2 units and run 3 units to the right.

Plot a point at $(3, 3)$.

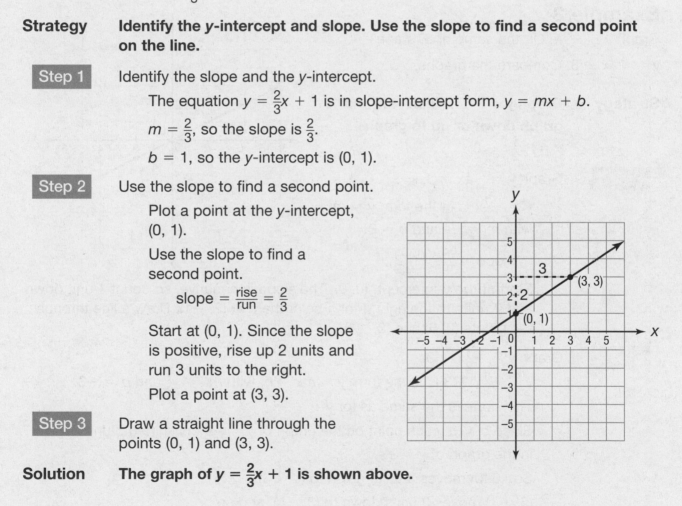

Step 3   Draw a straight line through the points $(0, 1)$ and $(3, 3)$.

Solution   **The graph of $y = \frac{2}{3}x + 1$ is shown above.**

A linear equation in the form $y = mx$ has $b = 0$. This means that its y-intercept is at $(0, 0)$, the **origin**.

> To graph $y = mx + b$:
>
> - First, graph $y = mx$.
>
> - Shift each point on the graph up or down b units.
>
>   If $b > 0$, shift the graph b units up.
>
>   If $b < 0$, shift the graph b units down.

Example 3

Graph $y = -\frac{1}{2}x$. On the same grid, graph $y = -\frac{1}{2}x - 3$. Compare the graphs.

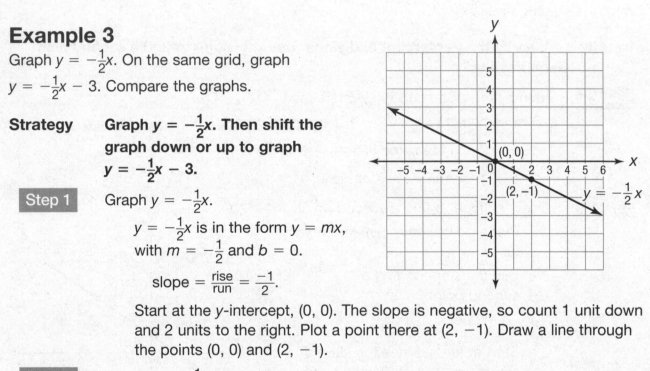

Strategy   **Graph $y = -\frac{1}{2}x$. Then shift the graph down or up to graph $y = -\frac{1}{2}x - 3$.**

Step 1   Graph $y = -\frac{1}{2}x$.

$y = -\frac{1}{2}x$ is in the form $y = mx$, with $m = -\frac{1}{2}$ and $b = 0$.

$$\text{slope} = \frac{\text{rise}}{\text{run}} = \frac{-1}{2}.$$

Start at the y-intercept, $(0, 0)$. The slope is negative, so count 1 unit down and 2 units to the right. Plot a point there at $(2, -1)$. Draw a line through the points $(0, 0)$ and $(2, -1)$.

Step 2   Graph $y = -\frac{1}{2}x - 3$.

$y = -\frac{1}{2}x - 3$ is in the form $y = mx + b$, with $m = -\frac{1}{2}$ and $b = -3$.
The slope is the same as for $y = -\frac{1}{2}x$.

Since $b < 0$, each point on the graph of $y = -\frac{1}{2}x$ is shifted 3 units down in the graph of $y = -\frac{1}{2}x - 3$.

So, $(0, 0)$ moves 3 units down to $(0, -3)$.

$(2, -1)$ moves 3 units down to $(2, -4)$, and so on.

Solution   The graphs of $y = -\frac{1}{2}x$ and $y = -\frac{1}{2}x - 3$ are shown on the right. Their slopes are the same. Every point in the graph of $y = -\frac{1}{2}x - 3$ is shifted three units down from the graph of $y = -\frac{1}{2}x$.

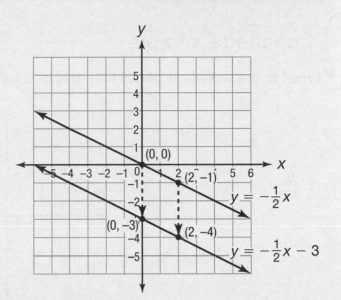

When you know the coordinates of one point on a line and its slope, you can use the **point-slope form** to write the equation in slope-intercept form.

> A line that passes through (x_1, y_1) with a slope m can be written in point-slope form:
>
> $$y - y_1 = m(x - x_1)$$

Example 4

A line has a slope of $\frac{4}{3}$ and passes through (3, 5). Write the equation of the line in slope-intercept form.

Strategy   Use the point-slope form to write the equation.

The slope, m, is $\frac{4}{3}$. Let $(x_1, y_1) = (3, 5)$.

$$y - y_1 = m(x - x_1)$$
$$y - 5 = \frac{4}{3}(x - 3)$$
$$y - 5 = \frac{4}{3}x - 4 \qquad \text{Distribute } \frac{4}{3} \text{ over } (x - 3).$$
$$y - 5 + 5 = \frac{4}{3}x - 4 + 5 \qquad \text{Add 5 to both sides.}$$
$$y = \frac{4}{3}x + 1$$

Solution   A line with a slope of $\frac{4}{3}$ that passes through (3, 5) has the equation $y = \frac{4}{3}x + 1$.

Coached Example

What is the equation of the line graphed below?

Find the values of m and b to write an equation in slope-intercept form.

The y-intercept is the point at which the graph crosses the ____-axis.

The y-intercept of this graph is (0, ____). So, b = _____.

Choose two points on the graph to find the slope, m.

Use the y-intercept, (0, ___), and the point (3, 0).

To move from the y-intercept to (3, 0), count _____ units up and _____ units to the right.

$$m = \frac{\text{rise}}{\text{run}} = \underline{\quad} = \underline{\qquad}$$

Since the line slants _____ from left to right, the slope is positive.

Substitute those values of m and b into $y = mx + b$.

$$y = \underline{\quad}x - \underline{\quad}$$

The equation of the line is y = _____.

Lesson Practice

Choose the correct answer.

1. What is the *y*-intercept of this line?

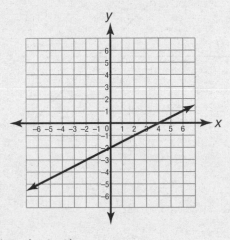

  A. $(-2, 0)$

  B. $(0, -2)$

  C. $(0, 4)$

  D. $(4, 0)$

2. What is the slope of the line whose equation is $y = \frac{1}{6}x - \frac{1}{2}$?

  A. $-\frac{1}{2}$ **C.** $\frac{1}{6}$

  B. $-\frac{1}{6}$ **D.** $\frac{1}{2}$

3. A line has a slope of -5 and passes through $(1, -1)$. Which is the equation of the line in point-slope form?

  A. $y - 1 = -5(x + 1)$

  B. $y - 5 = x - 1$

  C. $y + 1 = -5(x - 1)$

  D. $y + 1 = -5(x + 1)$

4. Which graph represents the equation $y = 2x - 1$?

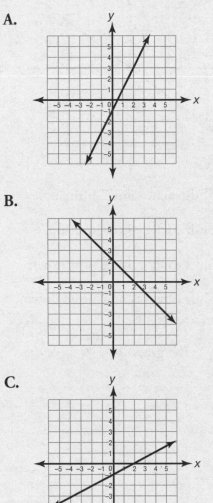

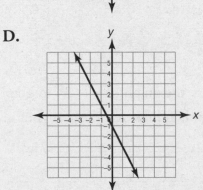

5. The graph below represents $y = \frac{3}{4}x$. Which describes how this graph would need to be shifted in order to graph $y = \frac{3}{4}x + 2$?

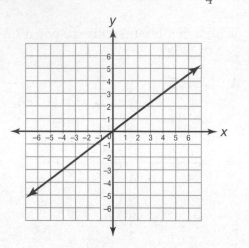

A. Shift each point 2 units up.

B. Shift each point $\frac{3}{4}$ unit up.

C. Shift each point $\frac{3}{4}$ unit down.

D. Shift each point 2 units down.

6. Which equation best represents the line graphed below?

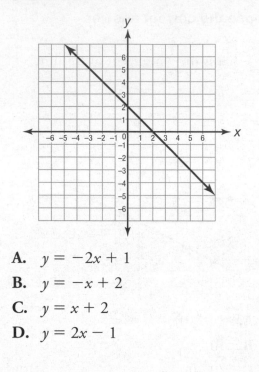

A. $y = -2x + 1$

B. $y = -x + 2$

C. $y = x + 2$

D. $y = 2x - 1$

7. Below is the graph of $y = 3x$.

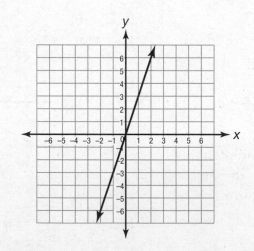

A. Use the graph of $y = 3x$ to graph $y = 3x - 4$ on the grid above.

B. Explain how you created your graph in Part A.

_____

_____

Proportional Relationships

Common Core State Standard:
8.EE.5

Getting the Idea

A ratio is a comparison of two numbers. For example, if there are 13 boys and 12 girls in a class, the ratio of boys to girls is 13 to 12. This can also be written with a colon, 13:12, or as a fraction, $\frac{13}{12}$. Since a ratio is a comparison of numbers, the ratio $\frac{13}{12}$ is not an improper fraction and cannot be rewritten as $1\frac{1}{12}$.

The ratio $\frac{13}{12}$ compares part of a class (the boys) to another part of the class (the girls). You can also use ratios to compare parts to totals.

Example 1

A bouquet contains only red and white roses. The ratio of red roses to white roses is 1:2. What is the ratio of red roses to the total number of roses in the bouquet?

Strategy   **Use the part-to-part ratio to find the part-to-total ratio.**

   Step 1   Write the part-to-part ratio as a fraction.

$$\frac{\text{red roses}}{\text{white roses}} = \frac{1}{2}$$

   Step 2   Use that ratio to write the part-to-total ratio.

The total includes all the red roses and all the white roses.

$$\frac{\text{red roses}}{\text{red roses} + \text{white roses}} = \frac{1}{1 + 2} = \frac{1}{3}$$

Solution   **The ratio of red roses to total roses is $\frac{1}{3}$.**

A **proportion** shows that two ratios are equal in value. You can use proportional reasoning to solve for an unknown value in a proportion.

Example 2

The debate team won 3 out of 5 debates it participated in this semester. If the team participated in 20 debates, how many debates did it win? How many did it lose?

Strategy   **Set up a proportion and use proportional reasoning.**

   Step 1   What does the given ratio represent?

$\frac{3}{5}$ is the ratio of debates won to total debates.

Step 2   Write a second ratio that includes the same terms.

Let x represent the number of debates won. There were 20 debates total.

$$\frac{\text{debates won}}{\text{total debates}} = \frac{x}{20}$$

Step 3   Set the ratios equal to each other to form a proportion.

$$\frac{3}{5} = \frac{x}{20}$$

Step 4   Use proportional reasoning and think about equivalent fractions to find the value of x.

The denominators are 5 and 20, and $5 \times 4 = 20$.

So, multiply the numerator and denominator of the first ratio by 4.

$$\frac{3}{5} = \frac{3 \times 4}{5 \times 4} = \frac{12}{20}$$

$\frac{3}{5}$ is equivalent to $\frac{12}{20}$, so the value of x is 12.

Step 5   Find the number of debates that the team lost.

If the team won 12 out of 20 debates, then the number of debates lost was: $20 - 12 = 8$.

Solution   **The team won 12 debates and lost 8 debates.**

Another way to solve for an unknown value in a proportion is to use cross-multiplication. To cross-multiply, multiply the numerator of each ratio by the denominator of the other ratio and set them equal to each other. Then solve for the unknown value.

A **rate** is a ratio that compares quantities that use different units. Use the same strategies to work with rates as you use with other ratios.

Example 3

It costs $261 for 3 nights at Pavia Pavilion Hotels. At the same rate, how much will it cost to stay for 7 nights?

Strategy   **Set up a proportion and cross-multiply.**

Step 1   Write a ratio comparing the cost to the number of nights.

$$\frac{\text{cost}}{\text{number of nights}} = \frac{261}{3}$$

Step 2   Write a second ratio that includes the unknown.

Let x represent the unknown cost.

$$\frac{\text{cost}}{\text{number of nights}} = \frac{x}{7}$$

Step 3   Set the ratios equal to each other to form a proportion.

$$\frac{261}{3} = \frac{x}{7}$$

Step 4   Cross-multiply and solve for x.

$$\frac{261}{3} = \frac{x}{7}$$

$$261 \cdot 7 = 3 \cdot x$$

$$1{,}827 = 3x$$

$$\frac{1{,}827}{3} = \frac{3x}{3}$$

$$609 = x$$

Solution   **A 7-night stay at the Pavia Pavilion Hotels will cost $609.**

A **unit rate** is a rate that, when expressed as a fraction, has a 1 in the denominator. For example, 46 miles per gallon is a unit rate because it expresses the ratio $\frac{46 \text{ miles}}{1 \text{ gallon}}$. If a unit rate involves money, it is called a **unit price**.

Example 4

A 16-ounce box of Wheaty Puffs costs $3.52. A 64-ounce box of Wheaty Puffs is sold at the same unit price. What is the cost of the 64-ounce box?

Strategy   **Find the unit price. Then use it to find the cost of the 64-ounce box.**

Step 1   Find the unit price of the 16-ounce box.

The rate is $\frac{\$3.52}{16 \text{ ounces}}$. So, divide by 16 to find the unit price.

$$\frac{3.52}{16} = \frac{3.52 \div 16}{16 \div 16} = \frac{0.22}{1}$$

The unit price is $0.22 per ounce.

Step 2   Multiply to find the cost of the 64-ounce box.

Since 1 ounce costs $0.22, multiply by 64 to find the price for 64 ounces.

$$\$0.22 \times 64 = \$14.08$$

Solution   **The cost of the 64-ounce box of Wheaty Puffs is $14.08.**

Coached Example

Mr. Cipriati has driven 700 miles in 4 days. If he continues to drive at the same rate, how many miles will he drive in 22 days?

Write two ratios, each comparing the number of miles to the number of days.

Let x represent the unknown quantity.

$$\frac{\text{miles}}{\text{days}} = \frac{700}{\rule{1cm}{0.4pt}} \qquad \frac{\text{miles}}{\text{days}} = \frac{x}{\rule{1cm}{0.4pt}}$$

Set the ratios equal to each other to form a proportion. Then cross-multiply to solve.

$$\frac{700}{\rule{2cm}{0.4pt}} = \frac{x}{\rule{2cm}{0.4pt}}$$

$$\rule{2cm}{0.4pt} \cdot \rule{2cm}{0.4pt} = 4 \cdot x \qquad \text{Cross-multiply.}$$

$$\rule{3cm}{0.4pt} = 4x$$

$$\rule{3cm}{0.4pt} = \frac{4x}{4} \qquad \text{Divide by 4.}$$

$$\rule{3cm}{0.4pt} = x$$

If he continues to drive at the same rate, Mr. Cipriati will drive _____ miles in 22 days.

Lesson Practice

Choose the correct answer.

1. A bag contains only cherry and grape gumballs. The ratio of cherry gumballs to grape gumballs in the bag is 7:6. What is the ratio of grape gumballs to total gumballs in the bag?

   A. 6:7

   B. 7:13

   C. 6:13

   D. 1:6

2. There are 30 seventh-grade students and 10 eighth-grade students in the school drama club. What is the ratio of eighth-grade students to seventh-grade students?

   A. 1 to 3

   B. 3 to 5

   C. 3 to 8

   D. 5 to 3

3. Millie can type 375 words in 5 minutes. If she types for 25 minutes at that rate, how many words will she have typed?

   A. 1,875

   B. 1,700

   C. 1,625

   D. 1,500

4. A major-league baseball team plays 162 games each season. So far this season, Rosanne's favorite team has won 17 games and lost 10 games. If the team continues to win at the same rate, how many games will Rosanne's favorite team win?

   A. 85

   B. 96

   C. 100

   D. 102

5. A box of 20 pencils costs $1.90. A box of 50 pencils has the same unit price as a box of 20 pencils. What is the cost of the box of 50 pencils?

   A. $0.76

   B. $3.80

   C. $4.25

   D. $4.75

6. Mr. Edelstein has driven 12 miles in 15 minutes. What is his speed in miles per hour?

   A. 36 miles per hour

   B. 44 miles per hour

   C. 48 miles per hour

   D. 60 miles per hour

7. There were $13\frac{1}{2}$ gallons of gas in Max's tank. He then used 8 gallons of gas to drive 208 miles. If he continues to use gas at the same rate, which proportion could be used to find m, the number of miles he can drive before his tank is empty?

A. $\frac{208}{8} = \frac{m}{5.5}$

B. $\frac{208}{8} = \frac{5.5}{m}$

C. $\frac{208}{8} = \frac{m}{13.5}$

D. $\frac{208}{8} = \frac{13.5}{m}$

8. Participants in a charity event can either walk or run 5 kilometers to raise money for charity. The ratio of runners to walkers this year is 5 to 12. There are a total of 120 runners this year. What is the total number of participants in the event?

A. 500

B. 408

C. 288

D. 24

9. The table shows the prices of three sizes of Miracle Sparkling Water.

Sparkling Water Prices

Size	Price
12 fl oz	?
32 fl oz	$2.24
64 fl oz	$5.12

A. The unit price for the 12-fluid-ounce (fl oz) size is $0.11 per fluid ounce. What is the price for buying 12 fluid ounces of Miracle Sparkling Water? Show your work.

B. Which size offers the better unit price—32 fluid ounces or 64 fluid ounces? Explain your reasoning and show your work.

_____

_____

_____

Direct Proportions

Common Core State Standard:

8.EE.5

Getting the Idea

A **direct proportion** is a special kind of linear equation. In a direct proportion, the ratio of two variables, such as y and x, is a constant, m. That means that for every change in x, y changes by a constant factor, m. We can say that y is directly proportional to x.

A direct proportion may be written in one of the following forms:

$$y = mx \text{ or } \frac{y}{x} = m$$

where $m \neq 0$ and m is the **constant of proportionality**, as well as the slope of the line that represents the equation.

Example 1

Which graph shows a direct proportion?

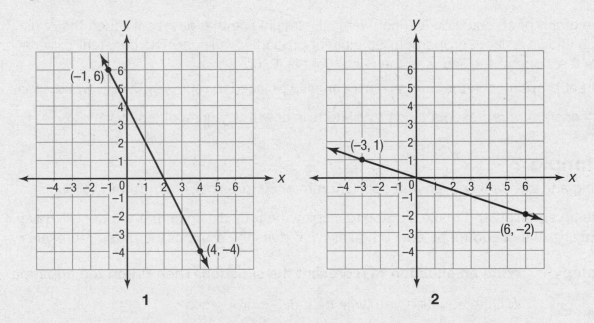

1

2

Strategy   Find the ratio of y to x for at least two points on each line. Determine if the ratio is constant.

Step 1 Find the ratio $\frac{y}{x}$ for two points on the line in Graph 1.

For $(-1, 6)$, the ratio is: $\frac{6}{-1} = -6$.

For $(4, -4)$, the ratio is: $\frac{-4}{4} = -1$. ➞ different ratio

The ratio $\frac{y}{x}$ is not constant. So, Graph 1 does not show a direct proportion.

Step 2 Find the ratio $\frac{y}{x}$ for two points on the line in Graph 2.

For $(-3, 1)$, the ratio is: $\frac{1}{-3} = -\frac{1}{3}$.

For $(6, -2)$, the ratio is: $\frac{-2}{6} = \frac{-1}{3} = -\frac{1}{3}$. ➞ same ratio

The ratio $\frac{y}{x}$ is constant for both points.

Since all points on a straight line change by the same constant rate, the slope, you only need to test those two points.

Graph 2 shows a direct proportion.

Solution **Graph 2 shows a direct proportion because the ratio of y to x, $-\frac{1}{3}$, is a constant.**

If the graph of an equation is a non-vertical, straight line that passes through the origin, the graph shows a direct proportion. So a direct proportion is linear. The constant of variation, m, is the slope of the line, and the y-intercept is $(0, 0)$.

A direct proportion in the form $y = mx$ can also be used to represent a real-world situation.

The constant m represents a unit rate and tells how many units of y per unit of x.

Example 2

During a trip, a car is driven at a constant rate of 60 miles per hour on the highway.

Write an equation and make a graph to display the total distance that the car will travel if it maintains that speed for at least 3 hours. What does the slope of the graph represent?

Strategy **Write an equation to represent the situation. Then graph the equation.**

Step 1 Will the equation you write be a direct proportion?

As the hours increase, the distance traveled increases.

The car travels at a constant rate.

So, the total distance traveled is directly proportional to the number of hours that the car is driven.

Step 2	Write an equation in the form $y = mx$.

Let x represent the number of hours.

Let y represent the total distance, in miles.

The car travels at a constant rate of speed: $\frac{60 \text{ miles}}{1 \text{ hour}}$. So, the constant, m, is 60.

The equation is $y = 60x$.

Step 3	Draw and label a coordinate grid.

The car cannot travel a negative number of miles or drive for a negative number of hours, so use only the 1st quadrant.

Title the graph and label its axes.

Step 4	Find two points and connect them with a line.

The graph is a direct proportion, so it must pass through (0, 0).

Since you must show at least 3 hours, substitute 3 for x in the equation to find another point.

$y = 60(3)$

$y = 180$

Plot a second point at (3, 180). Draw a line through the points.

Step 5	What does the slope of the graph represent?

In $y = 60x$, $m = 60$. So, the slope of the graph, 60, shows the speed of the car.

Solution **The equation $y = 60x$ and the graph in Step 4 represent this situation.**

The slope of the graph represents the speed of the car, 60 miles per hour.

A direct proportion can be represented using an equation, a graph, or a table. Sometimes, you may need to compare two different representations.

Example 3

Cassie has to buy several pounds of tomatoes at a farmer's market. The graph shows the cost of buying tomatoes at Farm Stand 1.

The equation $y = 4x$ gives the cost of buying x pounds of tomatoes at Farm Stand 2. Which farm stand offers the better price?

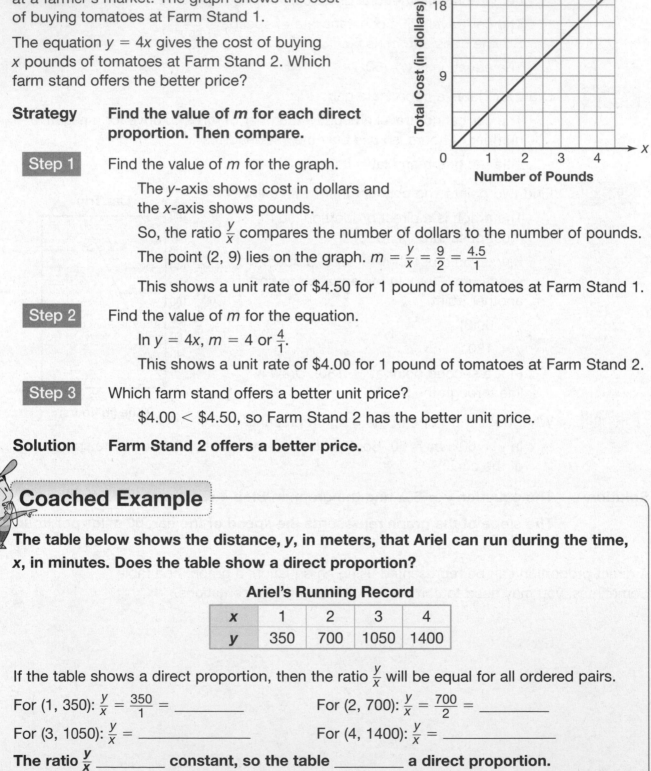

Tomatoes at Farm Stand 1

Strategy   **Find the value of m for each direct proportion. Then compare.**

Step 1   Find the value of m for the graph.

The y-axis shows cost in dollars and the x-axis shows pounds.

So, the ratio $\frac{y}{x}$ compares the number of dollars to the number of pounds.

The point (2, 9) lies on the graph. $m = \frac{y}{x} = \frac{9}{2} = \frac{4.5}{1}$

This shows a unit rate of $4.50 for 1 pound of tomatoes at Farm Stand 1.

Step 2   Find the value of m for the equation.

In $y = 4x$, $m = 4$ or $\frac{4}{1}$.

This shows a unit rate of $4.00 for 1 pound of tomatoes at Farm Stand 2.

Step 3   Which farm stand offers a better unit price?

$4.00 < $4.50, so Farm Stand 2 has the better unit price.

Solution   **Farm Stand 2 offers a better price.**

Coached Example

The table below shows the distance, y, in meters, that Ariel can run during the time, x, in minutes. Does the table show a direct proportion?

Ariel's Running Record

x	1	2	3	4
y	350	700	1050	1400

If the table shows a direct proportion, then the ratio $\frac{y}{x}$ will be equal for all ordered pairs.

For (1, 350): $\frac{y}{x} = \frac{350}{1} = $ _____   For (2, 700): $\frac{y}{x} = \frac{700}{2} = $ _____

For (3, 1050): $\frac{y}{x} = $ _____   For (4, 1400): $\frac{y}{x} = $ _____

The ratio $\frac{y}{x}$ _____ constant, so the table _____ a direct proportion.

Lesson Practice

Choose the correct answer.

1. Which equation represents a direct proportion?

   A. $y = x - 2$ C. $y = \frac{2}{x}$

   B. $y = x + 2$ D. $y = 2x$

2. Two trains, Train 1 and Train 2, are traveling at a constant rate of speed. The equation $y = 122x$ shows the total distance in miles, y, traveled by Train 1 over x hours. The graph below shows the relationship between the traveling time and the distance traveled for Train 2.

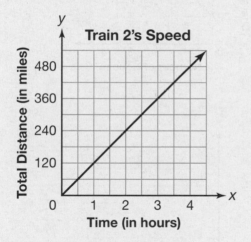

Which statement is true?

   A. Train 1 is traveling at a faster speed than Train 2.

   B. Train 2 is traveling at a faster speed than Train 1.

   C. Train 1 and Train 2 are traveling at the same speed.

   D. Train 1 and Train 2 are **not** traveling at a constant speed.

3. Sandra and Emil are both house painters, and each charges an hourly rate for a painting job. The equation $y = 13x$ shows the total charge, y, in dollars, for hiring Sandra to paint a house for x hours. The table below shows the same information for Emil.

Emil's Charges

x	2	4	6	8
y	26	52	78	104

Which statement is true?

   A. Sandra's hourly rate is $1.00 cheaper.

   B. Emil's hourly rate is $1.00 cheaper.

   C. Emil's hourly rate is $13.00 cheaper.

   D. Sandra and Emil work for the same hourly rate.

4. Which table represents a direct proportion?

   A.
x	1	2	3	4
y	2	5	10	17

   B.
x	1	2	3	4
y	2	3	4	5

   C.
x	1	2	3	4
y	0	1	2	2

   D.
x	1	2	3	4
y	4	8	12	16

5. Each graph shows the rate charged by four different landscapers for a landscaping job. Which graph shows a direct proportion?

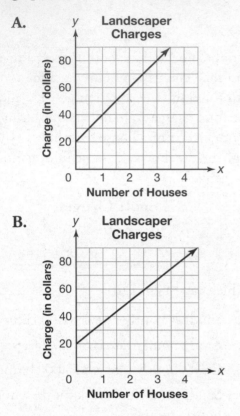

A.

Landscaper Charges

Charge (in dollars) vs. Number of Houses

C.

Landscaper Charges

Charge (in dollars) vs. Number of Houses

B.

Landscaper Charges

Charge (in dollars) vs. Number of Houses

D.

Landscaper Charges

Charge (in dollars) vs. Number of Houses

6. At Sal's Service Station, the total cost, y, in dollars, of buying x gallons of premium gasoline can be represented by the equation $y = 3x$.

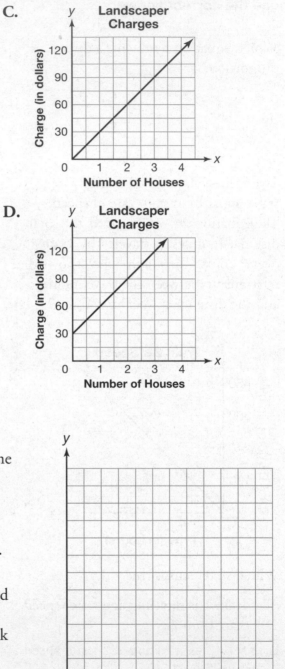

A. On the grid, make a graph to display the total cost of buying up to 10 gallons of premium gasoline at Sal's Service Station. Title your graph and label each axis.

B. What is the slope of the graph you created in Part A? What does the slope represent in the problem situation? Show your work and/or explain your reasoning.

_____

_____

_____

_____

Pairs of Linear Equations

Common Core State Standard:

8.EE.8

Getting the Idea

You can use what you know about the slope of a line to help you classify a pair of lines.

Type of Lines	Example
Intersecting lines cross one another at a point. They have different slopes. $y = -x$ has a slope of -1. $y = 2x - 3$ has a slope of 2.	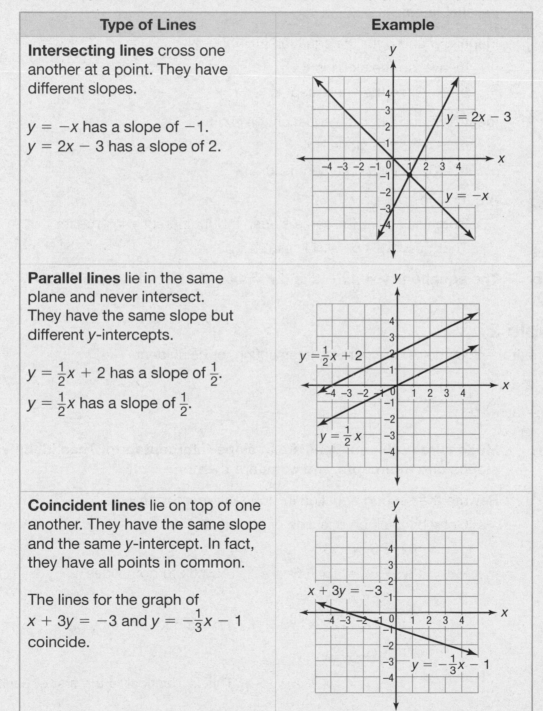
Parallel lines lie in the same plane and never intersect. They have the same slope but different y-intercepts. $y = \frac{1}{2}x + 2$ has a slope of $\frac{1}{2}$. $y = \frac{1}{2}x$ has a slope of $\frac{1}{2}$.	
Coincident lines lie on top of one another. They have the same slope and the same y-intercept. In fact, they have all points in common. The lines for the graph of $x + 3y = -3$ and $y = -\frac{1}{3}x - 1$ coincide.	

Example 1

Do these equations represent parallel lines?

$$y = 4x - 2$$
$$y = 4x + 3$$

Strategy   **Find the slope and y-intercept for each line. Then use the definition of parallel lines.**

Step 1   Identify m and b for the equation $y = 4x - 2$.

        $m = 4$, so the slope is 4.

        $b = -2$, so the y-intercept is $(0, -2)$.

Step 2   Identify m and b for the equation $y = 4x + 3$.

        $m = 4$, so the slope is 4.

        $b = 3$, so the y-intercept is $(0, 3)$.

Step 3   Are the lines parallel?

        Both lines have the same slope, 4, but different y-intercepts.

        That describes parallel lines.

Solution   **The equations $y = 4x - 2$ and $y = 4x + 3$ represent parallel lines.**

Example 2

Determine if the lines below are parallel, intersecting, or coincident.

$$y = \frac{3}{4}x + 2$$

$$-6x + 8y = 16$$

Strategy   **Make sure both equations are in slope-intercept form. Then identify the slopes and intercepts, and compare them.**

Step 1   Rewrite the second equation in slope-intercept form.

        Get y by itself on one side of the equation.

$$-6x + 8y = 16$$
$$-6x + 8y + 6x = 16 + 6x \qquad \text{Add } 6x \text{ to both sides.}$$
$$8y = 6x + 16$$
$$\frac{8y}{8} = \frac{6x}{8} + \frac{16}{8} \qquad \text{Divide both sides by 8.}$$
$$y = \frac{6}{8}x + 2$$
$$y = \frac{3}{4}x + 2 \quad \longrightarrow \quad \text{This is identical to the first equation.}$$

Step 2   Compare the slopes and *y*-intercepts.

The slope-intercept form of both equations is $y = \frac{3}{4}x + 2$.

So, both equations have the same slope, $\frac{3}{4}$, the same *y*-intercept, (0, 2), and share all points in common. They are coincident lines.

Solution   **The lines represented by $y = \frac{3}{4}x + 2$ and $-3x + 4y = 8$ are coincident lines.**

Example 3

Does the line that passes through (0, −3) and (2, 7) intersect the line that passes through (5, 1) and (10, 2)?

Strategy   **Find and compare the slopes.**

Step 1   Find the slope of the line that passes through (0, −3) and (2, 7).

$$m = \frac{y_2 - y_1}{x_2 - x_1}$$

$$m = \frac{7 - (-3)}{2 - 0} = \frac{10}{2}$$

$$m = 5$$

Step 2   Find the slope of the line that passes through (5, 1) and (10, 2).

$$m = \frac{y_2 - y_1}{x_2 - x_1}$$

$$m = \frac{2 - 1}{10 - 5}$$

$$m = \frac{1}{5} \qquad \longrightarrow \text{Different than a slope of 5.}$$

Step 3   Do the lines intersect?

The lines have different slopes, so they must intersect.

Solution   **The line that passes through (0, −3) and (2, 7) intersects the line that passes through (5, 1) and (10, 2).**

Coached Example

Does the line that passes through (0, 2) and (5, 5) intersect the line that passes through (−10, −4) and (−5, −1)? If not, are the two lines parallel or coincident?

Find the slope of the line that passes through (0, 2) and (5, 5).

$$m = \frac{y_2 - y_1}{x_2 - x_1} = \frac{5 - 2}{5} - 0 = \underline{\hspace{2cm}}$$

Find the slope of the line that passes through (−10, −4) and (−5, −1).

$$m = \frac{y_2 - y_1}{x_2 - x_1} = \underline{\hspace{1.5cm}} = \underline{\hspace{2cm}}$$

Are the slopes the same or different?

The slopes are _____. So, the lines are not intersecting lines.

To decide if the lines are parallel or coincident, compare their y-intercepts.

You know the first line passes through (0, 2). That is its y-intercept.

Use the point-slope form to find the y-intercept of the other line.

The slope, m, is _____. Let $(x_1, y_1) = (-5, -1)$.

$$y - y_1 = m(x - x_1)$$

$$y - (-1) = \left(\underline{\hspace{0.5cm}}\right)(x - \underline{\hspace{0.5cm}})$$

$$y + 1 = \underline{\hspace{0.3cm}}x + \underline{\hspace{0.5cm}}$$

$$y + 1 - 1 = \underline{\hspace{2cm}} - 1 \qquad \text{Subtract 1 from both sides.}$$

$$y = \underline{\hspace{2.5cm}}$$

The equation above is in slope-intercept form.

Since $b = $ _____, the y-intercept of that line is (0, _____).

Is that different or the same as the y-intercept of the first line? _____

**The lines have the same _____ and _____, so they _____ intersect.
They are _____ lines.**

Lesson Practice

Choose the correct answer.

1. Which describes the lines represented by these equations?

$$y = \frac{1}{6}x + 6$$
$$y = \frac{2}{12}x + 6$$

   A. coincident lines
   B. intersecting lines
   C. parallel lines
   D. vertical lines

2. Which describes the lines represented by these equations?

$$y = \frac{1}{2}x - 4$$
$$2y = x + 2$$

   A. coincident lines
   B. intersecting lines
   C. parallel lines
   D. vertical lines

3. Which could **not** be the slope of a line that intersects the line represented by $y = -x + 2$ at exactly one point?

   A. 1
   B. 0
   C. $-\frac{1}{2}$
   D. -1

4. Which equation represents a line that intersects $y = \frac{1}{3}x - 6$ at exactly one point?

   A. $y = \frac{1}{3}x + 4$
   B. $y = \frac{1}{3}x$
   C. $y = \frac{2}{4}x - 6$
   D. $y = \frac{2}{6}x - 6$

5. Which equation represents a line that is parallel but **not** coincident to $y = -\frac{1}{4}x + 1$?

   A. $y = \frac{1}{4}x + 2$
   B. $y = -\frac{2}{8}x + 2$
   C. $y = -\frac{3}{12}x + 1$
   D. $y = -4x + 1$

6. Which is true of the graphs of these two equations?

$$y = \frac{1}{2}x + 1$$
$$2y = x + 2$$

   A. The lines lie on top of one another.
   B. The lines intersect at exactly one point.
   C. The lines are parallel.
   D. The lines lie in the same plane, but they never intersect.

7. Which best describes the graphs of the line that has a slope of $-\frac{2}{5}$ and passes through $(10, -2)$ and the line that passes through $(0, 1)$ and $(5, -1)$?

   A. coincident lines

   B. intersecting lines

   C. parallel lines

   D. vertical lines

8. Which best describes the graphs of the line that passes through $(-4, 5)$ and $(0, 0)$ and the line that passes through $(4, 6)$ and $(12, 16)$?

   A. coincident lines

   B. intersecting lines

   C. parallel lines

   D. vertical lines

9. On a coordinate plane, Aiden draws a line that passes through $(0, 6)$ and $(1, 7)$ and another line that passes through $(-4, -6)$ and $(2, 0)$.

   A. Find the slope of each line, showing each step in the process.

   B. Are the lines that Aiden drew intersecting, parallel, or coincident? Show your work and explain how you determined your answer.

   _____

   _____

   _____

Common Core State Standards:
8.EE.8.a, 8.EE.8.b

Solve Systems of Equations Graphically

Getting the Idea

A **system of linear equations** is two or more linear equations with the same variables. You can use what you've learned about pairs of linear equations to solve systems of linear equations.

One way to solve a system of linear equations is to graph both equations. If the lines intersect, the ordered pair that names the point of intersection is the solution for the system of equations. Since both lines pass through that point of intersection, that ordered pair of values satisfies both equations at the same time.

A system of equations may have one unique solution, no solution, or infinitely many solutions, as shown below.

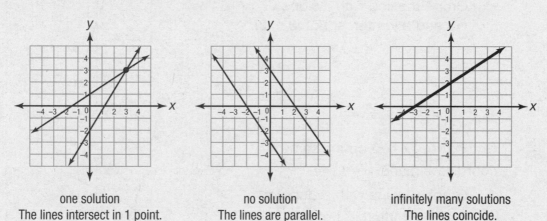

one solution	no solution	infinitely many solutions
The lines intersect in 1 point.	The lines are parallel.	The lines coincide.

Example 1

Solve the system of equations graphically.

$$3x + y = -2$$

$$y = \frac{1}{2}x + 5$$

Strategy **Graph each line. If the lines intersect, identify the coordinates of their point of intersection.**

Step 1 Graph $3x + y = -2$ on the coordinate plane.

Rewrite the equation in slope-intercept form.

$$3x + y = -2$$

$$3x + y - 3x = -3x - 2$$

$$y = -3x - 2$$

Graph the line with a slope of -3 and a y-intercept of $(0, -2)$.

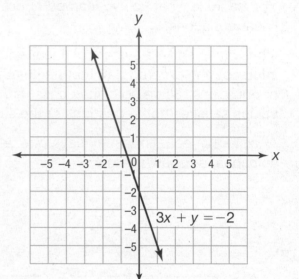

Step 2 Graph $y = \frac{1}{2}x + 5$ on the same coordinate plane.

Graph the line with a slope of $\frac{1}{2}$ and a y-intercept of $(0, 5)$.

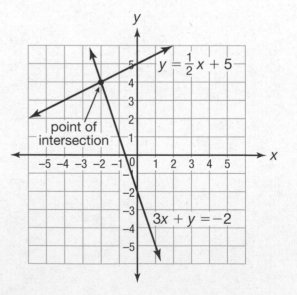

point of intersection

Step 3 Identify the solution for the system of linear equations.

The two lines intersect at $(-2, 4)$.

So, only when $x = -2$ and $y = 4$ are both equations true.

Step 4   Check the solution.

Substitute $(-2, 4)$ into each equation.

$$3x + y = -2 \qquad\qquad y = \tfrac{1}{2}x + 5$$

$$3(-2) + (4) \overset{?}{=} -2 \qquad\qquad 4 \overset{?}{=} \tfrac{1}{2}(-2) + 5$$

$$-6 + 4 \overset{?}{=} -2 \qquad\qquad 4 \overset{?}{=} -1 + 5$$

$$-2 = -2 \quad \checkmark \qquad\qquad 4 = 4 \quad \checkmark$$

Solution   **The solution for the system of equations is $(-2, 4)$.**

Example 2

Graph the system of equations below.

$$4x + 2y = 8$$
$$y = -2x - 1$$

How many solutions does this system have?

Strategy   **Graph each line. Then determine how many solutions the system has.**

Step 1   Graph $4x + 2y = 8$ on the coordinate plane.

Rewrite the equation in slope-intercept form.

$$4x + 2y = 8$$

$$2y = -4x + 8 \qquad\qquad \text{Subtract } 4x \text{ from both sides.}$$

$$y = -2x + 4 \qquad\qquad \text{Divide both sides by 2.}$$

Graph the line with a slope of -2 and a y-intercept of $(0, 4)$.

Step 2 Graph $y = -2x - 1$ on the coordinate plane.

Graph the line with a slope of -2 and a y-intercept of $(0, -1)$.

Step 3 Determine the number of solutions.

Both lines have the same slope, -2.

The lines are parallel, so there is no ordered pair that names a point on both lines.

The system has no solution.

Solution **The system of linear equations graphed above has no solution.**

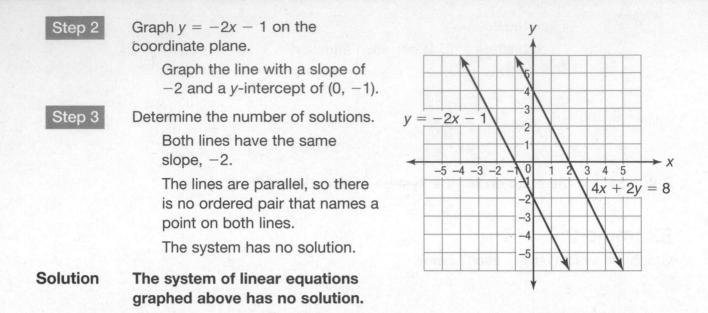

Example 3

Graph and estimate the solution for the system of equations below.

$$y = \tfrac{1}{2}x + 2$$
$$y = \tfrac{3}{2}x + 1$$

Strategy **Graph each line. If the lines intersect, find their point of intersection and visually estimate its coordinates.**

Step 1 Graph a line for each equation on the coordinate plane.

For $y = \tfrac{1}{2}x + 2$, graph a line with a slope of $\tfrac{1}{2}$ and a y-intercept of $(0, 2)$.

For $y = \tfrac{3}{2}x + 1$, graph a line with a slope of $\tfrac{3}{2}$ and a y-intercept of $(0, 1)$.

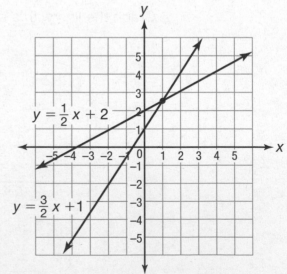

Step 2 Visually estimate the solution to the graph.

The lines appear to intersect when $x = 1$ and when the value of y is halfway between 2 and 3.

A good estimate of the solution is $(1, 2.5)$ or $\left(1, \tfrac{5}{2}\right)$.

Solution **A good estimate of the solution of the system of equations is $(1, 2.5)$.**

Coached Example

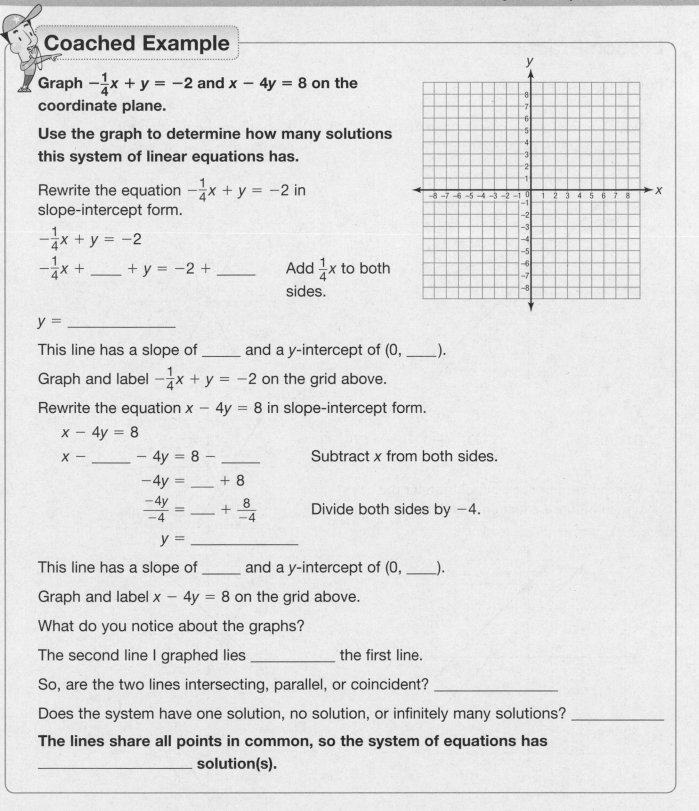

Graph $-\frac{1}{4}x + y = -2$ and $x - 4y = 8$ on the coordinate plane.

Use the graph to determine how many solutions this system of linear equations has.

Rewrite the equation $-\frac{1}{4}x + y = -2$ in slope-intercept form.

$-\frac{1}{4}x + y = -2$

$-\frac{1}{4}x + \underline{\quad} + y = -2 + \underline{\quad}$ Add $\frac{1}{4}x$ to both sides.

$y = \underline{\hspace{3cm}}$

This line has a slope of _____ and a y-intercept of (0, ___).

Graph and label $-\frac{1}{4}x + y = -2$ on the grid above.

Rewrite the equation $x - 4y = 8$ in slope-intercept form.

$x - 4y = 8$

$x - \underline{\quad} - 4y = 8 - \underline{\quad}$ Subtract x from both sides.

$-4y = \underline{\quad} + 8$

$\frac{-4y}{-4} = \underline{\quad} + \frac{8}{-4}$ Divide both sides by -4.

$y = \underline{\hspace{3cm}}$

This line has a slope of _____ and a y-intercept of (0, ___).

Graph and label $x - 4y = 8$ on the grid above.

What do you notice about the graphs?

The second line I graphed lies _____ the first line.

So, are the two lines intersecting, parallel, or coincident? _____

Does the system have one solution, no solution, or infinitely many solutions? _____

The lines share all points in common, so the system of equations has _____ solution(s).

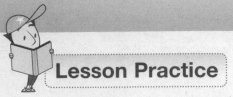

Lesson Practice

Choose the correct answer.

1. Which is the solution for the system of linear equations graphed below?

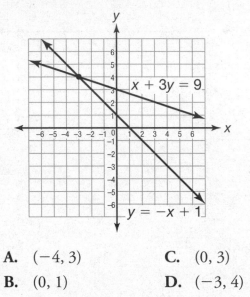

   A. $(-4, 3)$ **C.** $(0, 3)$

   B. $(0, 1)$ **D.** $(-3, 4)$

2. Which best describes the solution for the system of linear equations graphed below?

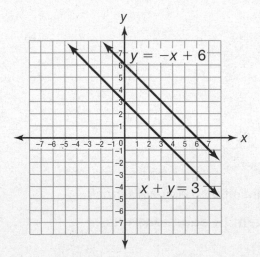

   A. $(3, 0)$ only

   B. $(6, 0)$ only

   C. no solution

   D. infinitely many solutions

3. Which shows the solution for the following system of equations?

   $$y = 2x$$
   $$2x + y = -4$$

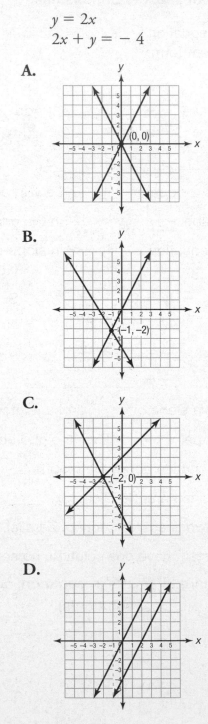

4. Which is the best estimate of the solution for the system of linear equations graphed below?

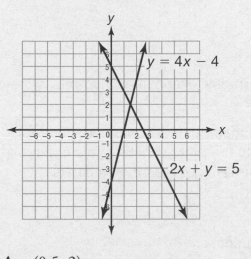

A. (0.5, 2)

B. (1.5, 2)

C. (2, 1.5)

D. (2.5, 1.5)

5. Solve this system of equations by graphing.

$$-2x + 5y = 10$$
$$y = x + 5$$

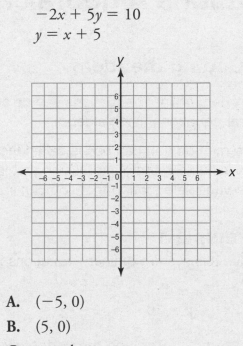

A. (−5, 0)

B. (5, 0)

C. no solution

D. infinitely many solutions

6. Consider the system of linear equations below.

$$3x + y = -4$$
$$-5x + y = 8$$

A. Graph the system of equations on the coordinate plane. Label each line on your graph and show any work you do.

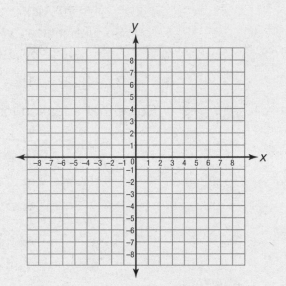

B. Use your graph from Part A to estimate the solution for the system of equations.

Explain how you determined your estimate.

_____

_____

_____

Common Core State Standards:
8.EE.8.a, 8.EE.8.b, 8.EE.8.c

Solve Systems of Equations Algebraically

Getting the Idea

A system of linear equations can be solved using graphs. You can also solve a system of linear equations using algebra.

One method you can use is called the substitution method. In this method, you replace one variable with an expression that is equivalent to the other variable. This creates a one-variable equation that you can solve to find the value of the other variable.

Example 1

Solve this system of linear equations using substitution.

$y = 2x - 1$
$2x + 2y = 10$

Strategy   **Replace the variable y using substitution.**

Step 1   Write an equation that contains only one variable.

The first equation is solved for y. It shows that the expression $2x - 1$ is equivalent to y.

So, substitute $2x - 1$ for y in the second equation.

$2x + 2y = 10$
$2x + 2(2x - 1) = 10$

Step 2   Solve that equation for x.

$2x + 2(2x - 1) = 10$	
$2x + 4x - 2 = 10$	Apply the distributive property.
$6x - 2 = 10$	Combine like terms.
$6x = 12$	Add 2 to both sides.
$\frac{6x}{6} = \frac{12}{6}$	Divide both sides by 6.
$x = 2$	

Step 3   Substitute that value for x into one of the original equations. Solve for y.

$y = 2x - 1$	
$y = 2(2) - 1$	Substitute 2 for x.
$y = 3$	

So, the solution is (2, 3).

Step 4	Check the solution.

Substitute (2, 3) into each equation.

$y = 2x - 1$	$2x + 2y = 10$
$3 \overset{?}{=} 2(2) - 1$	$2(2) + 2(3) \overset{?}{=} 10$
$3 \overset{?}{=} 4 - 1$	$4 + 6 \overset{?}{=} 10$
$3 = 3 \ \checkmark$	$10 = 10 \ \checkmark$

Solution   **The solution for this system of linear equations is (2, 3).**

Remember that a system of linear equations may have one unique solution, no solutions, or infinitely many solutions. You can use algebra to determine how many solutions a system has.

Example 2

Does this system of equations have one solution, no solution, or infinitely many solutions?

$$3x + 4y = -4$$

$$\tfrac{3}{4}x + y = -1$$

Strategy   **Solve the system using substitution.**

Step 1	Find an expression that is equivalent to one of the variables.

Solve the second equation for y.

$$\tfrac{3}{4}x + y = -1$$
$$y = -\tfrac{3}{4}x - 1$$

Step 2	Substitute that expression for y in the first equation.

$3x + 4y = -4$	
$3x + 4\left(-\tfrac{3}{4}x - 1\right) = -4$	Substitute $-\tfrac{3}{4}x - 1$ for y.
$3x + (-3x) + (-4) = -4$	Apply the distributive property.
$0x - 4 = -4$	Combine like terms.
$-4 = -4 \quad \longrightarrow$	always true

This means that this system has infinitely many solutions.

Note: If you write both equations in slope-intercept form, you will see that they each have the same slope, $-\tfrac{3}{4}$, and the same y-intercept, $(0, -1)$. So, they are coincident lines.

Solution   **This system of linear equations has infinitely many solutions.**

Another algebraic method you could use is called the elimination method. In this method, you add or subtract the equations to eliminate one variable and solve for the remaining variable. Then use that value to find the value of the other variable.

A **coefficient** is a number that is multiplied by a variable.

Example 3

Use elimination to determine if this system has one solution, no solution, or infinitely many solutions.

$$2x - 9y = 18$$
$$x + 3y = -21$$

Strategy   **Add the equations to eliminate one variable. Solve for the remaining variable and use its value to find the value of the other variable.**

Step 1   Eliminate one variable.

If you multiply the second equation by 3, the coefficients of y will be opposites (-9 and 9).

Their sum is zero, so that will eliminate y.

If you multiply the second equation by -2, the coefficients of x will be opposites (2 and -2).

Their sum is zero, so that will eliminate x.

It does not matter which variable you eliminate.

Step 2   Multiply the second equation by 3.

$$x + 3y = -21$$
$$3(x + 3y) = 3(-21)$$
$$3x + 9y = -63 \qquad \text{Apply the distributive property.}$$

Because you multiplied both sides of the equation by 3, the equation is still true.

Step 3   Add the result of Step 2 to the original first equation.

$$2x - 9y = 18$$
$$+ \ 3x + 9y = -63$$
$$\overline{5x + 0y = -45}$$
$$5x = -45$$

Step 4   Solve for x.

$$5x = -45$$
$$x = -9$$

Step 5 Substitute -9 for x into one of the original equations and solve for y.

$$x + 3y = -21$$
$$-9 + 3y = -21$$
$$3y = -12$$
$$y = -4$$

So, the solution is $(-9, -4)$.

Step 6 Check the solution, $(-9, -4)$, using each of the original equations.

$$2x - 9y = 18 \qquad\qquad x + 3y = -21$$
$$2(-9) - 9(-4) \overset{?}{=} 18 \qquad\qquad -9 + 3(-4) \overset{?}{=} -21$$
$$-18 + 36 \overset{?}{=} 18 \qquad\qquad -9 - 12 \overset{?}{=} -21$$
$$18 = 18 \quad \checkmark \qquad\qquad -21 = -21 \quad \checkmark$$

Solution **The solution for this system of linear equations is $(-9, -4)$.**

Sometimes, you can solve a system of linear equations using your reasoning skills.

Example 4

Does this system of equations have one solution, no solution, or infinitely many solutions?

$$5x - 4y = 2$$
$$5x - 4y = 3$$

Strategy **Since the equations look very similar, try to reason the answer.**

Step 1 Compare the equations.

Both equations have the same quantity, $5x - 4y$, on the left side.

Each equation has a different value on the right side.

Step 2 How many solutions do the equations have?

Since the quantity $5x - 4y$ cannot be equivalent to 2 and equivalent to 3, this system of equations has no solution.

Note: If you graphed the equations, you would see that they are parallel lines and therefore have no point of intersection. If you subtract the second equation from the first equation, you get the equation $0 = -1$, which is never true.

Solution **This system of equations has no solution.**

Example 5

A line passes through the points (1, −3) and (−4, 2). A second line passes through the points (1, 3) and (−2, −3). At what point do the two lines intersect?

Strategy   **Write a system of equations to represent the two pairs of points. Then solve the system using substitution.**

Step 1   Write the equation of the line that passes through the first pair of points.

Use the slope formula to find the slope of the line that passes through (1, −3) and (−4, 2).

$$m = \frac{y_2 - y_1}{x_2 - x_1} = \frac{2 - (-3)}{-4 - 1} = \frac{2 + 3}{-5} = \frac{5}{-5} = -1$$

Use the point (1, −3) and the value of m to write an equation in point-slope form.

$$y - y_1 = m(x - x_1)$$
$$y - (-3) = -1(x - 1)$$
$$y + 3 = -x + 1 \qquad \text{Apply the distributive property.}$$
$$y = -x - 2 \qquad \text{Subtract 3 from both sides to solve for } y.$$

Step 2   Write the equation of the line that passes through the second pair of points.

Use the slope formula to find the slope of the line that passes through (1, 3) and (−2, −3).

$$m = \frac{y_2 - y_1}{x_2 - x_1} = \frac{-3 - 3}{-2 - 1} = \frac{-6}{-3} = 2$$

Use the point (1, 3) and the value of m to write an equation in point-slope form.

$$y - y_1 = m(x - x_1)$$
$$y - 3 = 2(x - 1)$$
$$y - 3 = 2x - 2 \qquad \text{Apply the distributive property.}$$
$$y = 2x + 1 \qquad \text{Add 3 to both sides to solve for } y.$$

Step 3   Solve the system of equations using substitution.

Both equations are solved for y. The second equation shows that the expression $2x + 1$ is equivalent to y. Substitute $2x + 1$ for y in the first equation.

$$y = -x - 2$$
$$2x + 1 = -x - 2 \qquad \text{Substitute } 2x + 1 \text{ for } y.$$
$$3x + 1 = -2 \qquad \text{Add } x \text{ to both sides.}$$
$$3x = -3 \qquad \text{Subtract 1 from both sides.}$$
$$x = -1 \qquad \text{Divide both sides by 3.}$$

Step 4 Substitute the value of x into either equation to find the value of y. Use the first equation.

$$y = -x - 2 = -(-1) - 2 = 1 - 2 = -1$$

Step 5 Write the solution to the system of equations as an ordered pair.

$$x = -1, y = -1$$

So, the solution written as an ordered pair is $(-1, -1)$.

The solution to a system of two linear equations is the point of intersection.

Solution The lines intersect at $(-1, -1)$.

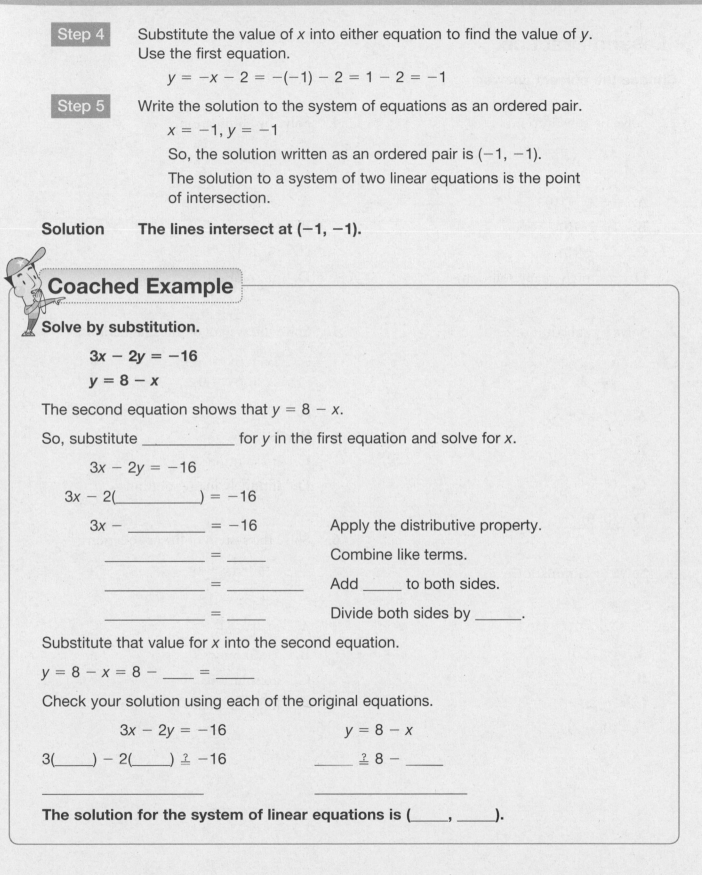

Coached Example

Solve by substitution.

$$3x - 2y = -16$$
$$y = 8 - x$$

The second equation shows that $y = 8 - x$.

So, substitute _____ for y in the first equation and solve for x.

$$3x - 2y = -16$$

$$3x - 2(\underline{\hspace{1.5cm}}) = -16$$

$$3x - \underline{\hspace{1.5cm}} = -16 \qquad \text{Apply the distributive property.}$$

$$\underline{\hspace{2cm}} = \underline{\hspace{1cm}} \qquad \text{Combine like terms.}$$

$$\underline{\hspace{2cm}} = \underline{\hspace{1.5cm}} \qquad \text{Add } \underline{\hspace{0.7cm}} \text{ to both sides.}$$

$$\underline{\hspace{2cm}} = \underline{\hspace{1cm}} \qquad \text{Divide both sides by } \underline{\hspace{0.9cm}}.$$

Substitute that value for x into the second equation.

$$y = 8 - x = 8 - \underline{\hspace{0.8cm}} = \underline{\hspace{1.2cm}}$$

Check your solution using each of the original equations.

$$3x - 2y = -16 \qquad\qquad y = 8 - x$$

$$3(\underline{\hspace{0.8cm}}) - 2(\underline{\hspace{0.8cm}}) \overset{?}{=} -16 \qquad\qquad \underline{\hspace{1cm}} \overset{?}{=} 8 - \underline{\hspace{1cm}}$$

$$\underline{\hspace{3cm}} \qquad\qquad\qquad \underline{\hspace{3cm}}$$

The solution for the system of linear equations is (__, __).

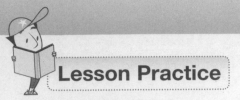

Lesson Practice

Choose the correct answer.

1. Solve by substitution

$$3x - 2y = 14$$
$$y = 5x$$

A. $(-2, -10)$

B. $(-2, 10)$

C. $(2, -10)$

D. infinitely many solutions

2. Solve by substitution.

$$x - 2y = -2$$
$$y = 2x + 4$$

A. $\left(-\frac{10}{3}, -\frac{8}{3}\right)$

B. $(-2, 0)$

C. $\left(-\frac{2}{3}, -\frac{4}{3}\right)$

D. $(0, 0)$

3. Solve by elimination.

$$x + 2y = -7$$
$$x - 5y = 7$$

A. $(-7, 0)$

B. $(-3, -2)$

C. $(-2, -3)$

D. $(0, -7)$

4. Solve by elimination.

$$4x - y = 1$$
$$x + 2y = 16$$

A. $(-2, -9)$

B. $(2, 7)$

C. $(3, 11)$

D. no solution

5. Solve the system of linear equations.

$$2x + 6y = -3$$
$$2x + 6y = 0$$

A. $(0, 2)$

B. $(9, 7)$

C. no solution

D. infinitely many solutions

6. Solve the system of linear equations.

$$y = -\frac{1}{3}x + 6$$
$$x + 3y = 18$$

A. $(2, 1)$ only

B. $(3, 6)$ only

C. no solution

D. infinitely many solutions

7. Solve the system of linear equations.

$$-3x + 5y = 4$$
$$x + y = -4$$

A. $(1, -5)$

B. $(0, -4)$

C. $(-3, -1)$

D. $(-5, 1)$

8. Solve the system of linear equations.

$$y = 6x + 7$$
$$3x - y = 2$$

A. $(-3, -11)$

B. $(-2, -8)$

C. $(-1, 1)$

D. no solution

9. Consider the system of linear equations shown below.

$$8x - 6y = -96$$
$$2x + 3y = 12$$

A. Solve this system using either the substitution or elimination method. Show your work or explain your answer.

B. Briefly describe how you can prove that the solution you found in Part A is correct. Show your work.

Common Core State Standard:
8.EE.8.c

Use Systems of Equations to Solve Problems

Getting the Idea

You can use a system of linear equations to model and solve problem situations in which you are given two different relationships between two unknown variables.

Example 1

Chelsea and Zack are both dog sitters. Chelsea charges $2 per day plus a sign-up fee of $3. Zack charges a flat rate of $3 per day. The system of linear equations below represents y, the total amount earned in dollars for x days of dog sitting.

$$y = 2x + 3$$
$$y = 3x$$

After how many days do Chelsea and Zack earn the same amount for dog sitting? What is that amount?

Strategy   **Graph a line for each equation.**

Step 1   Graph a line for each equation.

> For $y = 2x + 3$, graph a line with a slope of 2 and a y-intercept of (0, 3).
>
> Since Chelsea charges $2 per day plus a $3 sign-up fee, this line represents Chelsea's earnings.
>
> For $y = 3x$, graph a line with a slope of 3 and a y-intercept of (0, 0).
>
> Since Zack charges $3 per day, this line represents Zack's earnings.
>
> Note: Since Chelsea and Zack cannot work for a negative number of hours or earn a negative number of dollars, use only the 1st quadrant of a coordinate graph.

Dog Sitting Earnings

(graph showing Zack and Chelsea lines intersecting, Amount Earned (in dollars) on y-axis from 0 to 12, Number of Days on x-axis from 0 to 8)

Step 2   Identify and interpret the solution.

> The x-axis shows numbers of days. The y-axis shows amounts earned.
>
> The lines intersect at (3, 9).
>
> So, if Chelsea and Zack dog sit for 3 days, they each earn $9.

Solution   **Chelsea and Zack charge the same amount, $9, for 3 days of dog sitting.**

Example 2

Maria works for a gardener after school. Each week, she is paid an hourly rate plus a fixed amount to cover travel expenses. During the first week, Maria worked 10 hours and was paid $115. During the second week, Maria worked 5 hours and was paid $65. What is Maria's hourly rate of pay? What is the fixed amount she gets each week for travel expenses?

Strategy   **Write a system of linear equations. Use the elimination method to solve it.**

Step 1   Write a system of linear equations.

Let x represent Maria's hourly rate in dollars.

Let y represent the fixed amount for travel expenses.

During Week 1, she worked 10 hours and was paid $115. $10x + y = 115$

During Week 2, she worked 5 hours and was paid $65. $5x + y = 65$

Step 2   Multiply the second equation by -1.

This works because y and $-y$ are opposites.

$$5x + y = 65$$
$$-1(5x + y) = -1(65)$$
$$-5x - y = -65$$

Step 3   Add the result of Step 2 to the original first equation.

$$10x + y = 115$$
$$+ \ -5x - y = -65$$
$$\overline{5x = 50}$$

Step 4   Solve for x.

$$5x = 50$$
$$x = 10$$ ➜   This represents Maria's hourly rate of pay.

Step 5   Substitute 10 for x into one of the original equations and solve for y.

$$5x + y = 65$$
$$5(10) + y = 65$$
$$50 + y = 65$$
$$y = 15$$ ➜   This represents the fixed amount she is given for travel expenses.

Solution   **Maria earns $10 per hour and is paid a fixed amount of $15 each week for travel expenses.**

Coached Example

Inside the stables, there are only horses and people. The number of horses and people combined is 9. A boy inside the stables added the number of legs of all the horses and the number of legs of all the people combined. He determined that there are 30 legs in total. How many horses and how many people are in the stables?

Write a system of linear equations to represent this problem.

Let x represent the number of horses. Let y represent the number of people.

Remember that a horse has 4 legs and a person has 2 legs.

The number of horses and people combined is 9.   _____ $+ y = 9$

There are 30 legs in total.   _____ $+ 2y = 30$

Solve the first equation you wrote for y.

$y = $ _____

Substitute the equation you just rewrote for y into the second equation and solve for x.

_____ $+ 2($_____$) = 30$

_____ $= 30$ Apply the distributive property.

_____   Combine like terms.

_____   Subtract _____ from both sides.

_____   Divide both sides by _____.

Substitute that value for x into the first equation, $x + y = 9$.

____ $+ y = 9$

$y = $ _____   Subtract _____ from both sides.

The solution for the system of linear equations is (_____, _____).

It shows that there are _____ horses and _____ people in the stables.

Lesson Practice

Choose the correct answer.

1. 14 school festival tickets were sold to adults and children. A total of $38 was collected from these ticket sales. Adult tickets cost $4 each, and child tickets cost $1 each. The system of linear equations below represents x, the number of adult tickets sold, and y, the number of child tickets sold.

$$x + y = 14$$
$$4x + y = 38$$

How many tickets were sold?

A.   6 adult tickets and 8 child tickets

B.   8 adult tickets and 6 child tickets

C.   10 adult tickets and 4 child tickets

D.   12 adult tickets and 2 child tickets

2. Kaya and Tad started with the same number of baseball cards in their collections. Kaya collected 3 cards per week and now has 29 cards. Tad collected 2 cards per week and now has 20 cards. Let x represent the number of cards they began with, and let y represent the number of weeks. Which system of equations represents this situation?

A. $x + y = 20$
     $3x + 2y = 29$

B. $5y = 49$
     $x + 20 = 29$

C. $x + 3y = 20$
     $x + 2y = 29$

D. $x + 3y = 29$
     $x + 2y = 20$

3. The perimeter of an isosceles triangle is 16 inches. The length of its base is 2 times the length of one of its other sides. In the equations below, x represents the length of each of its equal sides, and y represents the length of its base.

$$2x + y = 16$$
$$y = 2x$$

What are the side lengths of the triangle?

A.   3 in., 3 in., 6 in.

B.   4 in., 4 in., 8 in.

C.   4 in., 8 in., 8 in.

D.   6 in., 6 in., 8 in.

4. Heidi paid $18 for 7 pairs of socks. She bought wool socks that cost $3 per pair and cotton socks that cost $2 per pair. How many pairs of socks did she buy?

A.   2 pairs of wool and 5 pairs of cotton

B.   3 pairs of wool and 4 pairs of cotton

C.   4 pairs of wool and 3 pairs of cotton

D.   5 pairs of wool and 2 pairs of cotton

5. A jar contains only dimes and nickels. The total number of coins in the jar is 15. The total value of the coins is $1.00. How many of each type of coin are in the jar?

A.   5 dimes and 10 nickels

B.   7 dimes and 8 nickels

C.   8 dimes and 7 nickels

D.   10 nickels and 5 dimes

6. Juan is trying to decide which parking garage to use. Bargain Garage charges a flat fee of $2 plus $4 per hour. Grey's Garage charges a flat fee of $8 plus $2 per hour. For how many hours will the cost of parking in either garage be the same, and what will that cost be?

A. For 2 hours, the cost at either garage will be $10.

B. For 2 hours, the cost at either garage will be $12.

C. For 3 hours, the cost at either garage will be $14.

D. For 3 hours, the cost at either garage will be $16.

7. James scored a total of 15 points during a basketball game. During the game, he made only free throws, worth 1 point each, and 3-point baskets. He made a total of 7 baskets. How many free throws and how many 3-point baskets did James make during the game?

A. 0 free throws and 7 three-point baskets

B. 1 free throw and 6 three-point baskets

C. 2 free throws and 5 three-point baskets

D. 3 free throws and 4 three-point baskets

8. Lucia wants to go ice skating. She must pay for admission and then rent ice skates. Rates for two rinks near her home are shown below.

Ice Plex	Skate World
$5 for admission plus $2 per hour for skate rental	$10 for admission plus $1 per hour for skate rental

A. Let x represent the cost of admission in dollars. Let y represent the number of hours during which a customer rents skates. Write a system of equations to represent this problem situation. Then graph the system on the coordinate grid.

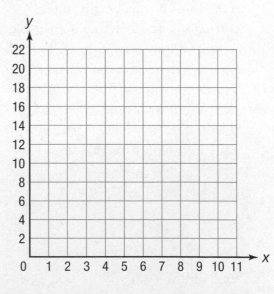

B. For how many hours of skate rental will the total cost be the same at both skating rinks? What would that total cost be? Use the solution for the system of linear equations to explain your answer.

Domain 2: Cumulative Assessment for Lessons 5–18

1. Which shows $7^{-2} \div 7^{-3}$ in exponential form?

   A. 7^{-5}

   B. 7^{-1}

   C. 7^{5}

   D. 7^{1}

2. To complete a specific task, a computer needs to carry out 6×10^{15} instructions. Keecia's computer has a 3.0 GHz microprocessor that can perform approximately 3×10^{9} instructions per second. Approximately how many seconds will it take for Keecia's computer to complete the task?

   A. 2×10^{6} seconds

   B. 2×10^{7} seconds

   C. 1.8×10^{7} seconds

   D. 1.8×10^{24} seconds

3. Which best describes the solution for this equation?

   $$2x - 6 = \tfrac{1}{2}(4x - 6)$$

   A. $x = -3$

   B. $x = -6$

   C. no solution

   D. infinitely many solutions

4. Calvin is running to train for a marathon. The graph below shows the total distance in kilometers, y, that he runs in x hours.

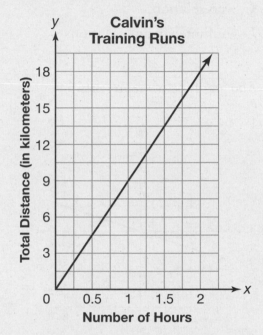

   What does the slope of the graph represent?

   A. Calvin's rate of speed, 4.5 kilometers per hour

   B. Calvin's rate of speed, 9 kilometers per hour

   C. the total distance that Calvin runs each time he trains, 18 kilometers

   D. the total number of hours that Calvin trains, 9 hours

5. Solve the system of linear equations.

$$2x - 3y = -6$$
$$y = 3x - 5$$

A. $(3, 4)$

B. $(-3, -14)$

C. no solution

D. infinitely many solutions

6. The graph below represents $y = \frac{1}{3}x$.

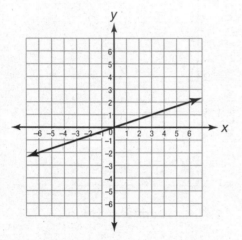

Which describes how this graph would need to be shifted in order to graph $y = \frac{1}{3}x + 4$?

A. Shift each point $\frac{1}{3}$ unit up.

B. Shift each point 4 units up.

C. Shift each point $\frac{1}{3}$ unit down.

D. Shift each point 4 units down.

7. Which equation represents a line that is parallel, but **not** coincident, to $y = \frac{4}{5}x + 10$?

A. $y = -\frac{4}{5}x - 10$

B. $y = \frac{8}{10}x + 10$

C. $y = \frac{12}{15}x + 5$

D. $y = -\frac{1}{2}x - 1$

8. Aisha and Gabriel both sew to earn extra money, and each charges an hourly rate for a job. The equation $y = 17.50x$ shows the total charge, y, in dollars for Aisha to do a sewing job. The table below shows the same information for Gabriel.

Gabriel's Charges

x	2	4	6	8
y	34	68	102	136

Which statement is true?

A. Gabriel's hourly rate is $0.50 cheaper.

B. Aisha's hourly rate is $0.50 cheaper.

C. Aisha's hourly rate is $16.50 cheaper.

D. Aisha and Gabriel work for the same hourly rate.

9. Solve for x.

$$x^3 = 729$$

10. Cecilia wants to rent a bike while on vacation. She also needs to rent a helmet. Rates for two bike rental shops are shown below.

Bernie's Bikes	Bike Land
$5 for the helmet plus $3 per hour for bike rental	free helmet rental plus $4 per hour

A. Let y represent the total cost of renting a bike and a helmet, in dollars.

Let x represent the number of hours that Cecilia rents a bike and a helmet.

Write a system of equations to represent this problem.

Then graph the system on the coordinate grid below.

_____

_____

B. For how many hours would Cecilia rent a bike and a helmet in order for the total cost to be the same at both shops? What would that total cost be?

Use the solution for the system of linear equations to explain your answer.

_____

_____

_____

Domain 3

Functions

Domain 3: Diagnostic Assessment for Lessons 19–23

Lesson 19 Introduction to Functions
8.F.1, 8.F.3

Lesson 20 Work with Linear Functions
8.F.3, 8.F.4

Lesson 21 Use Functions to Solve Problems
8.F.4

Lesson 22 Use Graphs to Describe Relationships
8.F.5

Lesson 23 Compare Relationships Represented in Different Ways
8.EE.5, 8.F.2

Domain 3: Cumulative Assessment for Lessons 19–23

Domain 3: Diagnostic Assessment for Lessons 19–23

1. Which mapping diagram represents a function?

A.
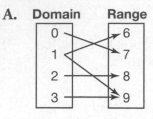

B.
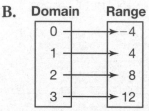

C.
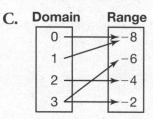

D.
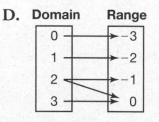

2. Given the linear function $y = -\frac{3}{2}x + 9$, what is the missing output value in the table below?

Input (x)	Output (y)
−8	21
−2	12
0	9
6	?

A. −3

B. 0

C. 2

D. 6

3. Sam went kayaking to a nearby island, stopped there for some lunch, and then kayaked back to the dock near his home. The graph shows the distance he traveled.

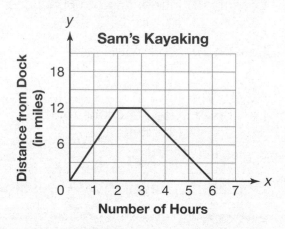

Which is the best description of what is shown by the graph?

A. Sam kayaked 12 miles in 2 hours, then stopped at an island for 1 hour, and finally spent 3 hours kayaking the 12 miles back to the dock.

B. Sam kayaked 12 miles in 2 hours, then stopped at an island for 1 hour, and finally kayaked 12 miles farther out to sea.

C. Sam kayaked 12 miles in 3 hours, then stopped at an island for 2 hours, and finally spent 1 hour kayaking back to the dock.

D. Sam kayaked 2 miles in 12 minutes, then stopped at an island for 1 hour, then kayaked 3 more miles in the next 12 minutes.

4. The table represents the linear function whose equation is $y = -10x + 80$.

Input (x)	Output (y)
0	80
2	60
4	40
6	20

What is the independent variable in this situation?

A. x

B. y

C. -10

D. 80

5. Which equation represents the linear relationship between x and y shown in the table?

x	0	1	2	3	4
y	3	1	-1	-3	-5

A. $y = -2x + 3$

B. $y = -2x + 1$

C. $y = -x + 1$

D. $y = 2x + 3$

6. What is the rate of change for the function represented by the table?

x	-2	-1	0	1	2
y	12	7	2	-3	-8

A. 5

B. $\frac{1}{5}$

C. $-\frac{1}{5}$

D. -5

7. Which graph does **not** represent a function?

A.

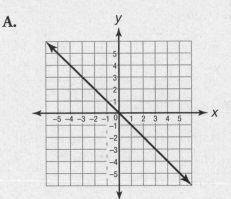

B.

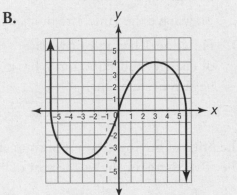

C.

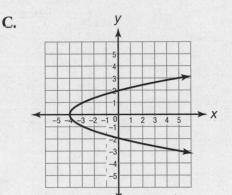

D.

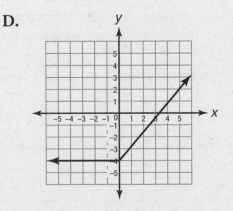

8. Function 1 is represented by the equation $y = 2x - 4$. The table below represents Function 2.

x	−1	0	1	2	3
y	1	3	5	7	9

Which statement is true?

A. Function 1 has a greater rate of change than Function 2.

B. Function 2 has a greater rate of change than Function 1.

C. They have the same rate of change.

D. Function 1 has a positive rate of change, and Function 2 has a negative rate of change.

9. A regular pentagon has all 5 sides congruent. The relationship between the side length of a regular pentagon, s, and its perimeter, P, can be modeled by the function $P = 5s$. Find the missing side length in the table below.

Side Length (s)	Perimeter (P)
1	5
3	15
8	40
?	65

10. The price of buying x bowls of soup at The Soup Spot is shown by the graph below. The advertisement shows the cost of buying x bowls of soup at The Lunch Box.

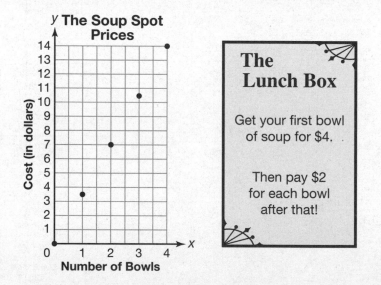

A. Determine the costs when a family buys 1, 2, 3, and 4 bowls of soup at each restaurant.

B. Explain when a family should buy soup at The Soup Spot and when they should buy soup at The Lunch Box. _____

_____

Introduction to Functions

Common Core State Standards:
8.F.1, 8.F.3

Getting the Idea

A **relation** is a set of ordered pairs. A **function** is a relation in which each input value, or x-value, corresponds to exactly one output value, or y-value. A function or other relation can be represented as a set of ordered pairs in a table, as an equation, or by a graph.

Example 1

Which table below represents a function?

Table 1

x	y
−2	−4
−1	−1
0	2
1	5

Table 2

x	y
4	−2
2	−1
1	0
2	1

Strategy   **Compare the x- and y-values.**

  Step 1   Compare the x- and y-values in Table 1.

The x-value −2 corresponds to only one y-value, −4.

The x-value −1 corresponds to only one y-value, −1.

The x-value 0 corresponds to only one y-value, 2.

The x-value 1 corresponds to only one y-value, 5.

Since each x-value has exactly one y-value, Table 1 shows a function.

  Step 2   Compare the x- and y-values in Table 2.

The x-value 4 corresponds to only one y-value, −2.

The x-value 2 corresponds to two y-values, −1 and 1.

Since there is an x-value that corresponds to more than one y-value, this relation is not a function.

Solution   **Table 1 represents a function. Table 2 represents a relation that is not a function.**

In a function, the set of all the input values, or *x*-values, is called the **domain**.

The set of all the output values, or *y*-values, is called the **range**.

Braces, { }, are often used when listing the domain and range.

Example 2

Identify the domain and range for the function shown below.

Sale Prices

Regular Price, *x*	$5	$10	$15	$20	$25
Sale Price, *y*	$1	$2	$3	$4	$5

Strategy   **Identify the domain and range of the function.**

Step 1   Identify the domain.

List the *x*-values.

5, 10, 15, 20, 25

Step 2   Identify the range.

List the *y*-values.

1, 2, 3, 4, 5

Solution   **The domain of the function is 5, 10, 15, 20, 25. The range is 1, 2, 3, 4, 5.**

Every function follows a **rule** that maps each element in its domain to exactly one element in its range. So, another way to determine if a relation is a function is to draw a mapping diagram. List the domain elements and the range elements in order. Then draw an arrow from each domain value to its range value. A mapping diagram for the function $y = x^2$ is shown below.

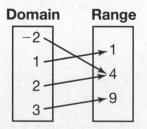

Notice that the *x*-values of −2 and 2 both map to 4, but there is still exactly one *y*-value for each *x*-value in the set. So, this relation is a function.

Example 3

Create a mapping diagram for the relation below.

(2, 6), (3, 9), (3, 12), (4, 15), (5, 10)

Is the relation a function?

Strategy	Create a mapping diagram.

Step 1　List the domain elements and the range elements in order.

List the domain values, 2, 3, 4, 5, in a box on the left.

List the range values, 6, 9, 10, 12, 15, in a box on the right.

Step 2　Draw an arrow from each domain value to its range value.

(2, 6) is part of the relation. So, draw an arrow from 2 to 6.

(3, 9) and (3, 12) are part of the relation. So, draw arrows from 3 to 9 and to 12.

Represent (4, 15) and (5, 10) with arrows, too.

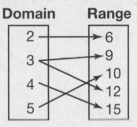

Step 3　Is the relation a function?

The domain element 3 maps to two different range elements, 9 and 12.

So, the relation is not a function.

Solution　**The mapping diagram in Step 2 shows that one of the domain elements maps to two range elements. So, the relation is <u>not</u> a function.**

The graph of a function is the set of ordered pairs consisting of input values and their corresponding output values. To determine whether a graph represents a function, you can use the **vertical line test**.

Imagine drawing vertical lines through the graph. If no vertical line intersects the graph in more than one point, the graph shows a function. For example, in the left-hand graph below, no vertical dashed line crosses the graph in more than one point, so the graph shows a function.

If you can draw a vertical line that intersects the graph in two or more points, the graph does not show a function. In the right-hand graph below, the vertical dashed line crosses the graph in two points, so the graph does not show a function.

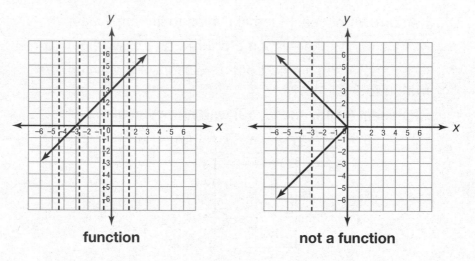

function   not a function

Example 4

Which graph represents a function?

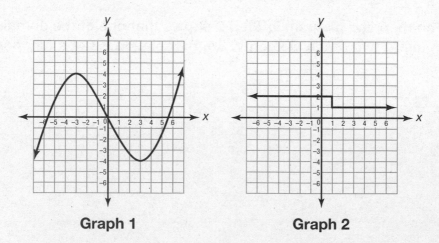

Graph 1   Graph 2

Strategy   **Use the vertical line test on each graph.**

Step 1   Use the vertical line test on Graph 1.

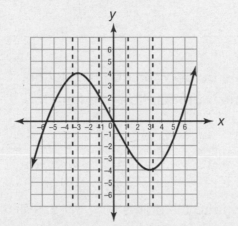

No matter where you draw a vertical line, it only crosses the graph once.

This graph represents a function.

Step 2   Use the vertical line test on Graph 2.

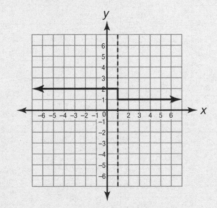

This graph includes a vertical segment at $x = 1$.

So, there is more than one y-value paired with the x-value, 1.

This graph does not represent a function.

Solution   **Graph 1 represents a function.**

Sometimes, the graph of a function or other relation is a set of connected points. Other times, the points are not connected.

Coached Example

The points shown in the table below represent a relation.
Plot the points and determine if the relation is a function.

x	1	2	2	3	4	6
y	−1	−2	−4	−4	−5	−6

The data in the table correspond to the points (1, −1), (2, −2), (2, ____), (____, ____), (____, ____), and (____, ____).

Plot those points on the coordinate grid below.

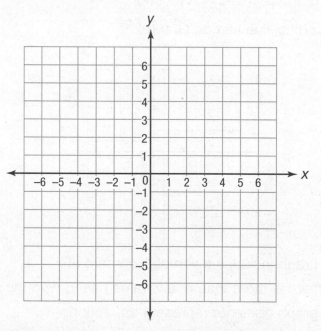

Draw a vertical line through (1, −1).

Does it pass through more than one point on the graph? _____

Draw a vertical line through (2, −1).

Does it pass through more than one point on the graph? _____

When x = 2, the relation corresponds to _____ y-value(s).

So, the relation _____ a function.

Lesson Practice

Choose the correct answer.

1. The table shows that the total amount charged, in dollars, by a hot dog vendor is a function of the number of hot dogs purchased.

Vendor Charges

Number of Hot Dogs, x	Total Charge, y
1	$2
2	$4
3	$6
4	$8
5	$10

What is the range of the function?

A. 5, 10

B. 1, 2, 3, 4, 5

C. 2, 4, 6, 8, 10

D. 1, 2, 3, 4, 5, 6, 8, 10

2. Which set of ordered pairs represents a function?

A. $(-2, 1), (0, 1), (1, -2), (3, 4)$

B. $(-1, 5), (-2, 3), (-2, 1), (-3, -1)$

C. $(12, 36), (9, 27), (-6, 30), (9, 18)$

D. $(3, 17), (-2, 11), (1, 8), (3, 5)$

3. Which table does **not** represent a function?

A.

x	7	8	8	9	10
y	7	14	21	28	35

B.

x	-2	-1	0	1	2
y	-8	-1	0	1	8

C.

x	-8	-4	0	4	8
y	2	0	-1	-3	-5

D.

x	-10	-5	0	5	10
y	5	5	5	5	5

4. Which graph does **not** represent a function?

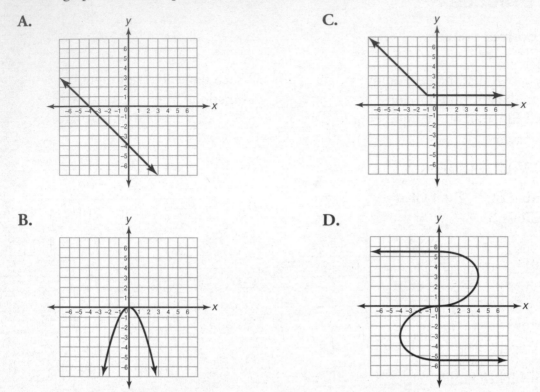

A.

C.

B.

D.

5. The table below shows a relation.

x	−5	−3	−1	0	1
y	−3	−6	0	−3	3

A. Identify the domain and range for the relation above. Then list the domain and range in the boxes below and create a mapping diagram for this relation.

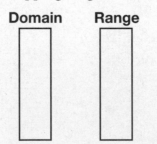

Domain   **Range**

B. Is the relation also a function? Use your mapping diagram to explain your answer.

_____

_____

_____

Work with Linear Functions

Common Core State Standards:
8.F.3, 8.F.4

Getting the Idea

A **linear function** is a special kind of function whose graph is a straight line. It can be represented by a linear equation in the form $y = mx + b$. A **nonlinear function** is any function that is not linear.

In a linear equation, no variable is raised to a power greater than 1. To review linear equations in that form, look back at Lesson 12.

Example 1

Is the function $y = x^3$ linear or nonlinear?

Strategy   **Look at the exponent for each variable in the equation.**

        In a linear equation, no variable is raised to a power greater than 1.

        In $y = x^3$, the variable x is raised to the power of 3.

        So, the equation is not a linear equation, and the function is nonlinear.

Solution   **The function represented by $y = x^3$ is a nonlinear function.**

You can also determine if a function is linear or nonlinear by graphing it. For example, the graph of the function from Example 1, $y = x^3$, is shown below. The points on the graph, $(-2, -8)$, $(-1, -1)$, $(0, 0)$, $(1, 1)$, and $(2, 8)$, do not lie on a straight line, so the function is nonlinear.

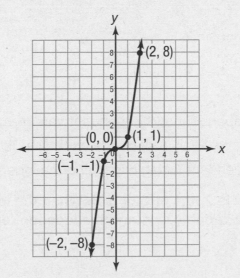

A function can describe a situation in which one quantity determines another. In a function, the output value, or y-value, depends on the input value, or x-value. That is why the y-value is often called the **dependent variable**, and the x-value is often called the **independent variable**.

Example 2

Given the linear function $y = -\frac{1}{2}x + 7$, find the missing output values in the table below. Then identify the independent and dependent variables.

Input (x)	Output (y)
−4	9
0	7
2	?
12	?

Strategy   **Substitute each x-value into the equation and solve for y. Then identify the independent and dependent variables.**

Step 1   Find the output when the input is 2.

$$y = -\frac{1}{2}x + 7$$

$$y = -\frac{1}{2}(2) + 7$$

$$y = -1 + 7$$

$$y = 6$$

Step 2   Find the output when the input is 12.

$$y = -\frac{1}{2}x + 7$$

$$y = -\frac{1}{2}(12) + 7$$

$$y = -6 + 7$$

$$y = 1$$

Step 3   Identify the independent and dependent variables.

The output, or y-value, depends on the input, or x-value.

So, the dependent variable is y, and the independent variable is x.

Solution   **When the input is 2, the output is 6. When the input is 12, the output is 1. The dependent variable is y, and the independent variable is x.**

A linear function can be represented in many ways. If you are given a table of values or a graph, you can use it to write an equation for the function. The equation for a function gives the rule that shows how each input value relates to each output value.

You can determine the **rate of change** and initial value for a linear function from a table of values, a graph, or an equation. The initial value is the value of y when x equals 0. In a graph, the rate of change is the same as the slope. Since the graph of a linear function is a straight line, the rate of change is constant, and you can determine it from any two pairs of (x, y) values for the function. You can find the initial value of a linear function by identifying the y-intercept.

Example 3

The input-output table below represents a linear function.

Input (x)	Output (y)
0	−1
1	2
2	5
3	8

Write an equation for the function and identify the rate of change and initial value.

Strategy   **Find the rule that relates each x-value to its corresponding y-value. Then write the equation.**

Step 1   Find a rule that relates the first pair of values in the table.

In the first column, the x-values are increasing by 1s.

In the second column, each y-value is 3 more than the previous y-value.

Look for a rule that involves multiplying by 3.

Consider $(0, -1)$: $0 \cdot \mathbf{3} = 0$, not -1. But, if you subtract 1, you get $0 - \mathbf{1} = -1$.

So, the rule may be: multiply each x-value by 3 and then subtract 1.

Step 2   See if the rule works for the other pairs of values in the table.

$1 \cdot \mathbf{3} - \mathbf{1} = 3 - 1 = 2$, and $(1, 2)$ is in the table.

$2 \cdot \mathbf{3} - \mathbf{1} = 6 - 1 = 5$, and $(2, 5)$ is in the table.

$3 \cdot \mathbf{3} - \mathbf{1} = 9 - 1 = 8$, and $(3, 8)$ is in the table.

Step 3   Use the rule to write an equation.

To find each y-value, multiply each x-value by 3 and then subtract 1.

So, $y = 3x - 1$.

Step 4 Determine the rate of change for the function.

Let $(x_1, y_1) = (1, 2)$.

Let $(x_2, y_2) = (2, 5)$.

rate of change $= \dfrac{y_2 - y_1}{x_2 - x_1} = \dfrac{5 - 2}{2 - 1} = \dfrac{3}{1} = 3$

Note: You could also have determined the rate of change by looking at the equation. In $y = 3x - 1$, $m = 3$. So, the slope of the graph and its rate of change must be 3.

Step 5 Determine the initial value.

The initial value is the value of y when x equals 0. The y-value $= -1$ when $x = 0$.

Solution The equation $y = 3x - 1$ describes the linear function. Its rate of change is 3, and the initial value is -1.

Coached Example

The graph represents a linear function. Find the rate of change for the function. Then write an equation for the function.

The rate of change for the function is equal to its

_____, m.

Choose two points on the graph to find the rate of change.

Let $(x_1, y_1) = (0, \underline{\hphantom{xx}})$.

Let $(x_2, y_2) = (2, \underline{\hphantom{xx}})$.

$m = \dfrac{y_2 - y_1}{x_2 - x_1} = \dfrac{\hphantom{xxx}}{2 - 0} = \underline{\hphantom{xx}}$

The equation for the line graphed above shows the equation of the linear function.

The y-intercept of the graph is $(0, \underline{\hphantom{xx}})$. That is the initial value. So, $b = \underline{\hphantom{xx}}$.

You already know that $m = \underline{\hphantom{xx}}$.

Substitute those values into the slope-intercept form, $y = mx + b$.

$y = \underline{\hphantom{x}}x + \underline{\hphantom{xx}}$

The rate of change for the linear function is _____, and its equation is $y = $ _____.

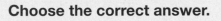

Lesson Practice

Choose the correct answer.

1. Which graph represents a linear function?

A.

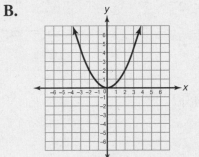

B.

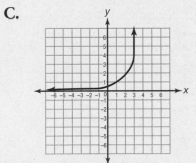

C.

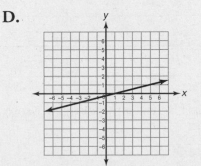

D.

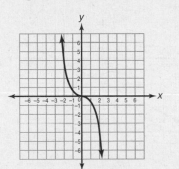

Use the table below for questions 2 and 3.

The table represents the function $y = 5x - 8$.

Input (x)	Output (y)
-2	-18
0	-8
1	?
?	12
8	?

2. What is the input when the output is 12?

A. 4

B. 5

C. 20

D. 52

3. Which statement is **not** true of the function?

A. The independent variable is x.

B. The dependent variable is y.

C. When the input is 1, the output is -2.

D. When the input is 8, the output is 32.

4. Which equation does **not** represent a linear function?

A. $y = \frac{1}{2}x + 2$

B. $y = x^2$

C. $y = 2x$

D. $y = x - 2$

5. Which equation represents the linear relationship between x and y shown in the table?

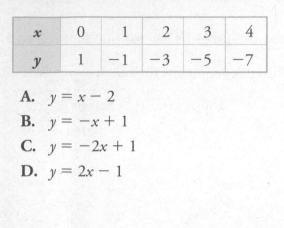

x	0	1	2	3	4
y	1	−1	−3	−5	−7

A. $y = x - 2$

B. $y = -x + 1$

C. $y = -2x + 1$

D. $y = 2x - 1$

6. Which equation represents the function graphed below?

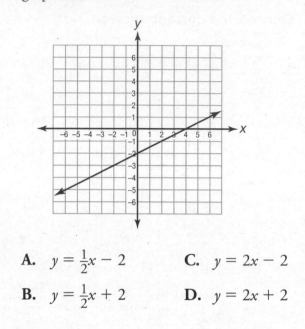

A. $y = \frac{1}{2}x - 2$ C. $y = 2x - 2$

B. $y = \frac{1}{2}x + 2$ D. $y = 2x + 2$

---

7. The ordered pairs in the table below represent a linear function.

Input (x)	Output (y)
0	6
1	10
2	14
3	18
...	...
?	50

A. Write an equation to represent the function shown above.

Explain how you determined your equation.

_____

_____

_____

B. Find the input for this function when the output is 50. Use the equation you wrote in Part A. Show your work.

Use Functions to Solve Problems

Common Core State Standard:
8.F.4

Getting the Idea

Sometimes, linear functions are used to model and solve real-world problems.

Example 1

The relationship between a side length of a square, s, and its perimeter, P, can be modeled by the function P = 4s. Find the missing perimeters in the table below. Then identify the dependent and independent variables.

Side Length (s)	Perimeter (P)
1	4
4	16
8	?
9.5	?

Strategy   **Substitute each s-value into the equation and solve for P. Then identify the independent and dependent variables.**

Step 1   Find the perimeter when the side length is 8 units.

$P = 4s$

$P = 4(8)$

$P = 32$

Step 2   Find the perimeter when the side length is 9.5 units.

$P = 4s$

$P = 4(9.5)$

$P = 38$

Step 3   Identify the independent and dependent variables.

The perimeter, P, depends on the length of a side of the square, s.

So, the dependent variable is P, and the independent variable is s.

Solution   **When a square has sides 8 units long, its perimeter is 32 units.**
When a square has sides 9.5 units long, its perimeter is 38 units.
In this function, the dependent variable is P, and the independent variable is s.

A graph may also be used to represent a real-world situation that is modeled by a linear function. If so, the slope of the graph shows the rate at which the two quantities in the problem are changing.

Example 2

The graph below shows how the cost of buying gas at Steve's Service Station changes, based on the number of gallons that are purchased.

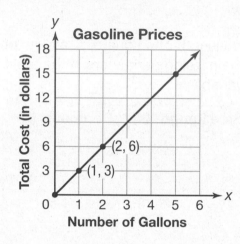

Find the price per gallon of gas.

Strategy   **Find and interpret the rate of change.**

Step 1   Find the rate of change.

This is the same as the slope of the graph.

Let $(x_1, y_1) = (1, 3)$.

Let $(x_2, y_2) = (2, 6)$.

rate of change $= \frac{y_2 - y_1}{x_2 - x_1} = \frac{6 - 3}{2 - 1} = \frac{3}{1} = 3$

Step 2   Interpret the rate of change.

Since the x-axis shows number of gallons and the y-axis shows total cost in dollars, the rate of change is $\frac{3 \text{ dollars}}{1 \text{ gallon}}$, or $3 per gallon.

Solution   **The gas costs $3 per gallon.**

Example 3

To bowl at Cavanaugh Lanes, it costs $2 per game plus a $3 shoe rental. The total cost, y, in dollars, depends on x, the number of games played. Write an equation to represent this situation. Then make a table of values to represent the situation.

Strategy   **Write an equation for the situation. Then make a table of values.**

Step 1   Translate the words into an equation.

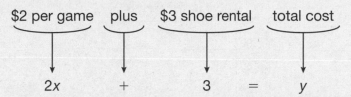

$2 per game	plus	$3 shoe rental	total cost
2x	+	3	= y

Note: This is a linear equation. The problem situation can be modeled by a linear function.

Step 2   Decide which x-values you should include in the table.

In real life, you cannot bowl fewer than 0 games. Also, if you go to the trouble of renting bowling shoes, you will probably bowl at least 1 game.

So, the initial x-value should be 1. The other x-values should be whole numbers because you can only pay for a whole number of games. In other words, the domain is limited to 1, 2, 3, 4, ….

Step 3   Make a table of values.

Games Bowled (x)	$y = 2x + 3$	Total Cost in Dollars (y)
1	$y = 2(1) + 3 = 5$	5
2	$y = 2(2) + 3 = 7$	7
3	$y = 2(3) + 3 = 9$	9
4	$y = 2(4) + 3 = 11$	11

Note: This table could be continued for other whole number x-values. However, when we make tables to represent functions, we accept that they usually only represent part of the function.

Solution   **The equation $y = 2x + 3$ and the table of values in Step 3 represent this problem situation.**

Example 4

A marching band needs to raise $1,200 in order to attend the regional marching band festival. The band members are selling tickets to a fundraising breakfast. Tickets are $10 for adults and $6 for children. The equation $10x + 6y = 1,200$, where x represents the number of adult tickets sold and y represents the number of children's tickets sold, can be used to model the situation. The graph of this equation is shown below.

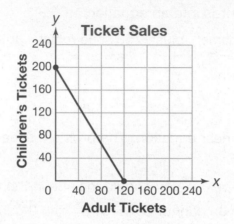

Explain the meaning of the y-intercept in terms of the number of adult and children's tickets sold.

Strategy   **Identify and interpret the y-intercept.**

Step 1   Identify the y-intercept.

It is located at (0, 200).

Step 2   Interpret the y-intercept.

Since y represents the number of children's tickets sold, the coordinates of the y-intercept, (0, 200), represent a situation where no adult tickets are sold (0 tickets) and only children's tickets are sold (200 tickets).

Solution   **The y-intercept represents a situation in which the marching band sells 200 children's tickets and 0 adult tickets to raise $1,200.**

Example 5

To rent a limousine from Deluxe Limousines, a customer must pay a set fee plus an additional amount per hour, as shown by the graph below.

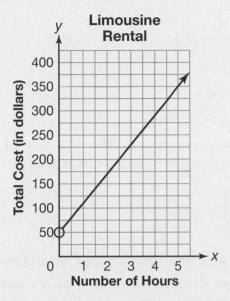

a. Identify and interpret the initial value and the rate of change.

b. Determine the cost of renting a limousine for 8 hours.

Strategy   **Use the graph and the slope formula to identify and interpret the y-intercept and the rate of change.**

Step 1	Find and interpret the initial value.

The initial value is the value of y when x is equal to 0. It is the y-intercept.

The y-intercept would be at (0, 50), but the dot at that point is open.

The open dot means that (0, 50) is not part of the solution.

This makes sense because no one would rent a limousine for 0 hours and pay $50.

However, the fact that the cost is $50 if the limousine is rented for 0 hours indicates that $50 is the set fee for a limousine rental before any hourly charges are added, or the initial value.

Step 2	Find and interpret the rate of change.

Let $(x_1, y_1) = (0, 50)$.

Let $(x_2, y_2) = (5, 350)$.

rate of change $= \dfrac{y_2 - y_1}{x_2 - x_1} = \dfrac{350 - 50}{5 - 0} = \dfrac{300}{5} = \dfrac{60}{1}$

Since the x-axis shows number of hours and the y-axis shows total cost in dollars, the rate of change is $\dfrac{\$60}{1 \text{ hour}}$, or $60 per hour.

This is the hourly rate charged for a limousine rental.

Step 3	Determine the cost of renting a limousine for 8 hours.

The x-axis of the graph does not show $x = 8$, so write an equation.

The cost, y, is a $50 fee plus $60 per hour for x hours, so:

$50 + 60x = y$

Substitute 8 for x and find the value of y.

$50 + 60(8) = y$

$50 + 480 = y$

$530 = y$

Solution   **The initial value, or set fee, is $50. The rate of change is $60 per hour. Renting a limousine for 8 hours costs $530.**

Coached Example

Students have dining cards at a boarding school. Each time a student gets a meal at the dining hall, 5 points are deducted from his or her dining card. Tyeisha's dining card had a value of 630 points at the beginning of the semester. If her card now has 570 points left on it, how many meals has she eaten at the dining hall?

Translate the words into an equation.

Let x represent the number of meals she has eaten.

Let y represent the total amount left on the card.

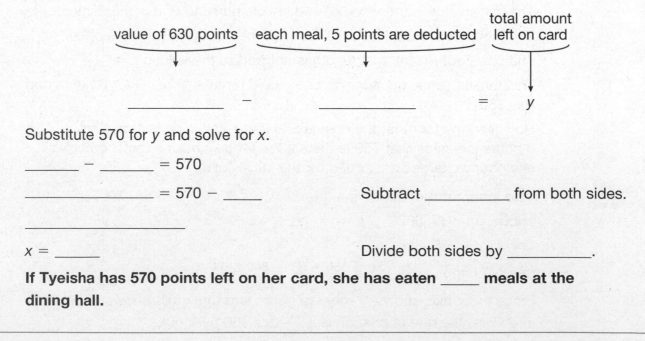

value of 630 points   each meal, 5 points are deducted   total amount left on card

_____ − _____ = y

Substitute 570 for y and solve for x.

_____ − _____ = 570

_____ = 570 − _____   Subtract _____ from both sides.

_____

$x =$ _____   Divide both sides by _____.

If Tyeisha has 570 points left on her card, she has eaten _____ meals at the dining hall.

Lesson Practice

Choose the correct answer.

Use the table for questions 1 and 2.

An equilateral triangle has 3 sides equal in length. The relationship between a side length of an equilateral triangle, s, and its perimeter, P, can be modeled by the function $P = 3s$.

Side Length (s)	Perimeter (P)
1	3
3	9
8	?
?	48

1. What is the perimeter of an equilateral triangle whose side length is 8 units?

   A. 4 units

   B. 11 units

   C. 12 units

   D. 24 units

2. What is the side length of an equilateral triangle whose perimeter is 48 units?

   A. 10 units

   B. 13 units

   C. 16 units

   D. 144 units

3. Ashton earns extra money by doing odd jobs for his neighbors. He charges a flat fee of $15 plus $8 per hour for each job. If he earned $47 for a job he did last week, how many hours did he work?

   A. 2   **C.** 3

   B. 2.6   **D.** 4

4. An online store sells T-shirts for $11 each. The store charges a $9 shipping and handling fee no matter how many shirts a customer orders. Which equation best represents y, the total cost in dollars, of buying x T-shirts from this online store?

   A. $11 + 9 + x = y$

   B. $11 + 9x = y$

   C. $11x + 9 = y$

   D. $11x + 9x = y$

5. Mrs. Ames uses a QuickPass device to pay bridge tolls when she drives. Each time she crosses a local toll bridge, $2 is automatically deducted from her QuickPass account. At the beginning of the month, she had a balance of $60 on her QuickPass account. If her current balance is $24, how many times has she crossed the bridge since the beginning of the month?

   A. 12

   B. 18

   C. 36

   D. 108

Use the graph for questions 6 and 7.

An accountant charges a set fee to complete a client's tax return, plus an additional rate for each hour she works on the return, as shown by the graph below.

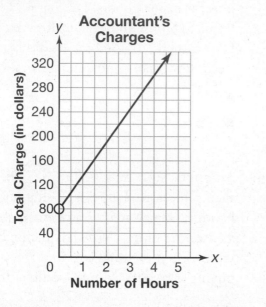

6. What does the slope of the graph represent?

   A. the accountant's hourly rate, $55 per hour

   B. the accountant's hourly rate, $75 per hour

   C. the set fee charged to complete a return, $80

   D. the set fee to complete any tax return, $300

7. What is the initial value?

   A. $0

   B. $80

   C. $140

   D. $160

8. For field trips, a museum charges a flat fee plus an additional rate for each student.

   The museum uses the equation of a linear function to determine y, the total cost in dollars, if x students attend the field trip. The table below shows a partial representation of this function.

Museum Field Trip Costs

Number of Students (x)	1	2	3	4	5
Total Cost in Dollars (y)	56	62	68	74	80

A. What is the rate of change shown by the table and what does it represent in the problem? Show your work and explain your answer.

_____

_____

B. The initial x-value in the table is 1. Explain why it is more appropriate for the table to start at $x = 1$ instead of $x = 0$.

_____

_____

Common Core State Standard:
8.F.5

Use Graphs to Describe Relationships

Getting the Idea

To represent some real-world situations, you may need to break a graph into pieces to show a sequence of events. It may not be as simple as drawing or interpreting a straight line.

Example 1

On Thursday, Maksim went for a long nature walk, stopping for lunch at one point.

The graph below represents his walk.

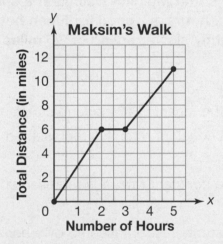

Describe what Maksim did during each interval shown.

Strategy   **Look at the graph piece by piece.**

Step 1   Look at the first piece of the graph.

The first piece is a line segment slanting up from (0, 0) to (2, 6).

Since the x-axis shows time, in hours, and the y-axis shows total distance, in miles, this segment shows that Maksim walked 6 miles during the first 2 hours.

Find and interpret the rate of change for this segment.

$$\text{rate of change} = \frac{\text{miles}}{\text{hours}} = \frac{6 - 0}{2 - 0} = \frac{6}{2} = 3$$

So, Maksim walked at a speed of 3 miles per hour for the first 2 hours.

Step 2 Look at the second piece of the graph.

   The second piece is a horizontal line segment from (2, 6) to (3, 6).

   The rate of change for a horizontal segment is 0.

   So, Maksim walked no additional distance during that hour.

   That is probably when he stopped for lunch.

Step 3 Look at the third piece of the graph.

   The third piece is a line segment slanting up from (3, 6) to (5, 11).

   $11 - 6 = 5$, so Maksim walked 5 more miles during the last 2 hours of his walk.

   $$\text{rate of change} = \frac{\text{miles}}{\text{hours}} = \frac{11 - 6}{5 - 3} = \frac{5}{2} = 2.5$$

   So, Maksim walked at a speed of 2.5 miles per hour for the last 2 hours.

Solution **The graph shows that Maksim walked at a speed of 3 miles per hour during the first 2 hours, stopped for lunch between hours 2 and 3, and walked at a slightly slower speed of 2.5 miles per hour between hours 3 and 5.**

Be careful when you interpret the meaning of the term *constant*.

- If a graph is increasing at a constant rate, it is represented by a line segment that slants up.

- If a graph is decreasing at a constant rate, it is represented by a line segment that slants down.

- If a piece of a graph is constant, it is represented by a horizontal line segment.

   Here, *constant* means that it is neither increasing nor decreasing.

Example 2

A function decreases at a constant rate from $(-6, 5)$ to $(-1, -3)$. It then increases at a constant rate from $(-1, -3)$ to $(2, 1)$. Finally, it is constant from $x = 2$ to $x = 4$. Sketch the graph.

Strategy   **Graph the function piece by piece.**

Step 1   Graph the first piece.

Plot points at $(-6, 5)$ and $(-1, -3)$ on a coordinate grid.

Since the graph decreases at a constant rate, draw a line segment to connect the points.

Step 2   Graph the second piece.

Plot a point at $(2, 1)$.

Since the graph increases at a constant rate, draw a line segment to connect $(-1, -3)$ to $(2, 1)$.

Step 3   Graph the final piece.

Since the graph is constant from $x = 2$ to $x = 4$, it will neither increase nor decrease during that interval.

$(2, 1)$ is on the graph, so plot a point with an *x*-value of 4 and the same *y*-coordinate as $(2, 1)$. That point is $(4, 1)$. Connect the two points with a horizontal line segment.

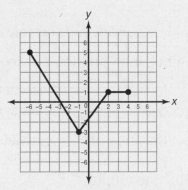

Solution   **The graph in Step 3 fits the verbal description of the function.**

Coached Example

The graph below shows Mr. Kowalski's commute home. He used a combination of taking the bus and walking to get home.

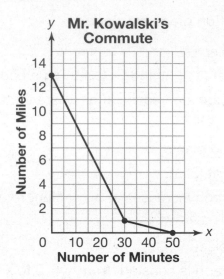

Mr. Kowalski's Commute

Use that information to describe each part of the graph.

Decide which line segment is steeper.

The line segment from (0, 13) to (30, 1) is _____ than the line segment from (30, 1) to (50, 0).

The steeper line segment shows Mr. Kowalski moving toward home at a faster rate.

Since a person travels faster on a bus than on foot, the line segment from (0, 13) to (30, 1) shows that Mr. Kowalski _____.

30 − 0 = 30, so that segment represents _____ minute(s) of his commute.

13 − 1 = _____, so it represents a distance of _____ mile(s) traveled.

Look at the line segment from (30, 1) to (50, 0).

Does this line segment show Mr. Kowalski taking the bus or walking? _____

50 − 30 = _____, so that segment represents _____ minute(s) of his commute.

1 − 0 = 1, so it represents a distance of _____ mile(s) traveled.

The graph shows that during his commute, Mr. Kowalski _____ for the first _____ minute(s) and traveled a distance of _____ mile(s). He then _____ for the next _____ minute(s) and traveled a distance of _____ mile(s).

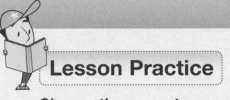

Lesson Practice

Choose the correct answer.

1. For which *x*-values is the interval increasing?

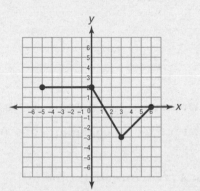

A. from $x = -5$ to $x = 0$

B. from $x = 0$ to $x = 3$

C. from $x = 3$ to $x = 6$

D. The graph does not show an increasing interval.

2. The graph shows the distance that Po-Ting drove during a road trip. At one point during the trip, Po-Ting took a break at a rest stop. During which time period did that occur?

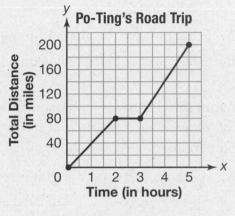

A. during the first 2 hours

B. between hours 2 and 3

C. between hours 3 and 5

D. between hours 4 and 5

3. Cara went roller skating. The graph shows the distance she traveled to and from her home.

Which best describes what is shown by the graph?

A. Cara skated up a hill for 20 minutes, then along the flat top of the hill for 10 minutes, and then down the hill for 30 minutes.

B. Cara skated down a hill for 20 minutes, then along the flat top of the hill for 10 minutes, and then up the hill for 30 minutes.

C. Cara skated toward her home for 20 minutes, then stayed at her home for 10 minutes, and then skated away from her home for 30 minutes.

D. Cara skated away from her home for 20 minutes, then took a break for 10 minutes, and then skated toward her home for 30 minutes.

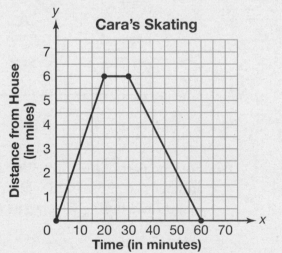

4. Which statement is true of the interval from $x = -4$ to $x = -1$?

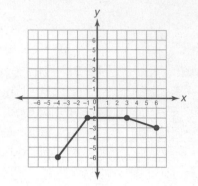

   A. That piece of the graph is nonlinear.   **C.** That piece of the graph is decreasing.

   B. That piece of the graph is linear.   **D.** That piece of the graph is constant.

5. Aaron traveled to his grandfather's house. He traveled at a constant rate for the first 20 minutes, for a distance of 1 mile. Then he stopped for 20 minutes to have a snack. Finally, he traveled at a constant rate over the next 40 minutes for a distance of 8 miles.

   A. Create a graph on the grid below to show Aaron's trip to his grandfather's house.

   B. For one part of his trip (either before or after his snack), Aaron rode his bicycle. For another part, he walked while pushing his bicycle. During which part did he ride his bicycle and during which part did he walk? Explain how you determined your answer.

_____

_____

_____

Common Core State Standards:
8.EE.5, 8.F.2

Compare Relationships Represented in Different Ways

Getting the Idea

You already know that functions can be represented in different ways—as a set of ordered pairs, in a table, with a verbal or algebraic rule, as an equation, or by a graph. Sometimes, you may need to compare two different functions represented in different ways.

Example 1

A brother and sister are racing 30 meters to a tree. Since Justin is younger, his sister Cami lets him have a 6-meter head start. The graph below shows the distance that Justin runs during the race.

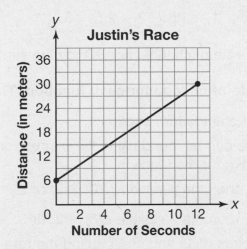

The equation $y = 3x$ can be used to represent y, the total distance in meters that Cami has run after x seconds have passed. Who is running at a faster speed? How much faster?

Strategy   Compare the rate of change shown in the graph to the rate of change shown by the equation.

Step 1   Find Justin's speed.

The x-axis shows time, in seconds, and the y-axis shows distance, in meters.

So, the rate of change for the graph compares seconds to meters.

Use the points (0, 6) and (12, 30) to find the rate of change.

$$\text{rate of change} = \frac{\text{meters}}{\text{seconds}} = \frac{30 - 6}{12 - 0} = \frac{24}{12} = 2$$

So, Justin's speed is 2 meters per second.

Step 2   Find Cami's speed.

The equation $y = 3x$ is in the form $y = mx$.

Since $m = 3$, the rate of change is 3 meters per second.

Step 3   Find the person running at a faster speed. How much faster?

3 meters per second > 2 meters per second, and $3 - 2 = 1$.

So, Cami is running 1 meter per second faster than Justin.

Solution   **Cami's speed is 1 meter per second faster than Justin's speed.**

Example 2

Compare the rates of change for the two linear functions described below.

Function 1: Any y-value can be found using the expression $\frac{1}{3}x + 2$.

Function 2:

x	−4	−2	0	2	4
y	1	1.5	2	2.5	3

Which function has a greater rate of change, or are they the same?

Strategy   **Identify the rate of change for each function. Then compare them.**

Step 1   Find the rate of change for Function 1.

Any y-value can be found using the expression $\frac{1}{3}x + 2$, so find two ordered pairs for the function.

If $x = 0$, then the y-value is $\frac{1}{3}(0) + 2$, or 2.

If $x = 3$, then the y-value is $\frac{1}{3}(3) + 2$, or 3.

Use the points (0, 2) and (3, 3) to find the rate of change.

$$\text{rate of change} = \frac{3 - 2}{3 - 0} = \frac{1}{3}$$

Note: You might notice that this function could also be represented as $y = \frac{1}{3}x + 2$. In that case, you can see that the rate of change is $\frac{1}{3}$ because $m = \frac{1}{3}$.

Step 2   Find the rate of change for Function 2.

Use the points (0, 2) and (4, 3) to find the rate of change.

$$\text{rate of change} = \frac{3 - 2}{4 - 0} = \frac{1}{4}$$

Step 3   Compare the rates of change.

$\frac{1}{3} > \frac{1}{4}$, so Function 1 has the greater rate of change.

Solution   **The rate of change for Function 1 is greater than for Function 2.**

Example 3

The price of buying x baseball caps at Sporty's is shown by the graph below. The advertisement shows the cost of buying the same baseball caps at Hats 'R Us.

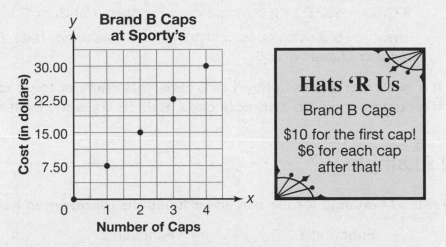

When does it make sense to buy caps at Sporty's? at Hats 'R Us?

Strategy   **List ordered pairs for each store and compare them.**

Step 1   List ordered pairs for each store.

It may help to list the pairs side by side in a table.

The ordered pairs for Sporty's can be found by looking at the graph.

To find the ordered pairs for Hats 'R Us, use the words in the advertisement.

The cost is $10 for the first cap and $6 for each cap after that.

Number of Caps (x)	Price at Sporty's (y)	Price at Hats 'R Us (y)
1	$7.50	$10.00
2	$15.00	$10 + $6 = $16.00
3	$22.50	$10 + $6(2) = $22.00
4	$30.00	$10 + $6(3) = $28.00

Compare the *y*-values, or prices, in the table.

$7.50 < $10.00, so buying 1 cap is cheaper at Sporty's.

$15.00 < $16.00, so buying 2 caps is also cheaper at Sporty's.

$22.50 > $22.00, so buying 3 caps is cheaper at Hats 'R Us.

The price of buying caps continues to be cheaper at Hats 'R Us for any number of caps over 3.

Solution   **If a customer wants to buy 1 or 2 caps, it is cheaper to go to Sporty's. If a customer wants 3 or more caps, Hats 'R Us is the better choice.**

Coached Example

Compare the rates of change for the two linear functions represented below.

Function 1

x	y
−4	−12
−2	−9
0	−6
2	−3
4	0

Function 2

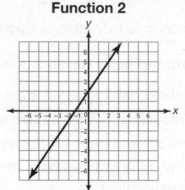

Which function has a greater rate of change, or are they the same?

Find the rate of change for Function 1.

Use the points (0, ____) and (____, 0).

$$m = \frac{0 - \underline{\quad}}{\underline{\quad} - 0} = \underline{\qquad}$$

Find the rate of change for Function 2.

Use the points (____, 2) and (____, 5).

$$\frac{m = 5 - 2}{\underline{\quad} - \underline{\quad}} = \underline{\quad}$$

Which function has a greater rate of change, or are they the same? _____

Function 1 has a rate of change of _____ and Function 2 has a rate of change of

_____, so _____.

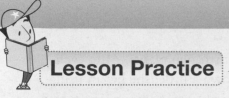

Lesson Practice

Choose the correct answer.

1. Compare the rates of change for the two functions represented below.

   For Function 1, any *y*-value can be found using the rule: multiply *x* by −2 and then add 2.

   The graph below represents Function 2.

   Function 2

   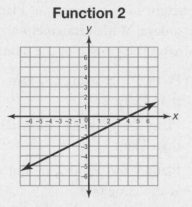

   Which statement about the rates of change for the two functions is true?

   A. Function 1 has a greater rate of change than Function 2.

   B. Function 1 and Function 2 have the same rate of change.

   C. Function 1 has a negative rate of change, and Function 2 has a positive rate of change.

   D. Function 1 has a positive rate of change, and Function 2 has a negative rate of change.

2. Compare the rates of change for the two linear functions represented below.

   Function 1

   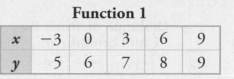

x	−3	0	3	6	9
y	5	6	7	8	9

   Function 2

   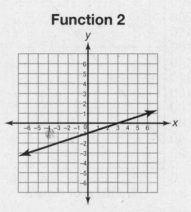

   Which statement about the rates of change for the two functions is true?

   A. Function 1 has a greater rate of change than Function 2.

   B. Function 2 has a greater rate of change than Function 1.

   C. Function 1 and Function 2 have the same rate of change.

   D. Function 2 does not have a constant rate of change, but Function 1 does.

3. Two trucks are driving on the same highway at the same time. Truck 1 is 10 miles from Smithtown. Its distance from Smithtown, in miles, after x hours can be found using the expression $10 + 55x$. The distance of Truck 2 from Smithtown is represented by the graph below.

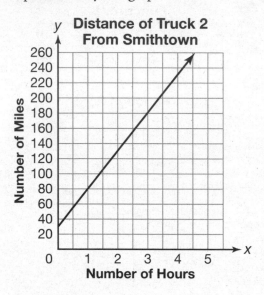

Exactly how many hours will pass before the two trucks are each the same distance from Smithtown?

A. 5   C. 3

B. 4   D. 2

4. Michelle planted two plants. After each plant had grown a little, she began using them for a science experiment. The table below shows the growth of Plant 1 over several days.

Number of Days (x)	0	1	2	3	4
Height in Centimeters (y)	1.5	3.5	5.5	7.5	9.5

The equation $y = 3 + 1.5x$ represents y, the height in centimeters, of Plant 2 over x days. Which statement accurately compares the growth of the plants?

A. Plant 1 is growing at a faster rate than Plant 2.

B. Plant 2 is growing at a faster rate than Plant 1.

C. Plant 1 and Plant 2 are both growing at the same rate.

D. Plant 2 was growing at a faster rate than Plant 1 at first, but then Plant 1 began to grow at a faster rate.

5. Two brothers, Leo and William, are racing their bikes around a 9.5-kilometer loop. Since Leo is younger, he is given a 1.5-kilometer head start. The graph below shows Leo's progress.

The expression $19x$ can be used to determine y, the distance in kilometers that William has traveled after x hours have passed. Which statement about this situation is **not** true?

A. William is traveling at a faster speed than Leo.

B. The difference between their speeds is exactly 1.5 kilometers per hour.

C. Leo traveled only 8 kilometers during the race.

D. Both brothers will finish the race in $\frac{1}{2}$ hour.

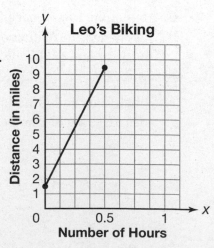

6. The price of buying *x* pairs of Brand B socks at Shoeway is shown by the graph below.

The advertisement shows the cost of buying several pairs of the same brand of socks at Foot World.

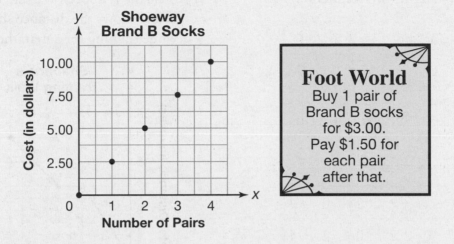

A. Determine the costs of buying 1, 2, 3, and 4 pairs of socks at each store. Show your work.

B. Explain when a customer should buy socks at Shoeway and when a customer should buy socks at Foot World.

_____

_____

_____

Domain 3: Cumulative Assessment for Lessons 19–23

1. Which table represents a function?

    A.
x	3	4	5	6	6
y	4	4	5	5	7

    B.
x	−2	−1	0	0	0
y	1	2	3	4	5

    C.
x	−4	−2	−2	0	2
y	0	2	0	−2	−4

    D.
x	−6	−5	−3	0	1
y	3	0	3	0	3

2. Given the linear function $y = -\frac{3}{2}x + 9$, what is the missing output value in the table below?

Input (x)	Output (y)
−8	21
−2	12
0	9
6	?

    A. −3
    B. 0
    C. 2
    D. 6

3. Shabana is a speed walker. The graph below shows the distance she walked while training for a marathon.

    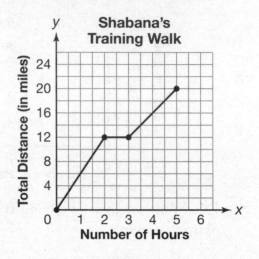

    During which part of the training walk did Shabana walk at the fastest speed?

    A. during the first two hours
    B. between hours 2 and 3
    C. between hours 3 and 5
    D. She walked at a constant rate of speed for the entire training walk.

4. The table represents the linear function whose equation is $c = 15 + 20x$. It shows the total cost, y, of renting a wallpaper hanger from a home improvement store for x days.

Number of Days (x)	Total Charge in Dollars (y)
1	35
2	55
3	75
4	95

What is the dependent variable in this situation?

A. 15   **C.** x

B. 20   **D.** y

5. Which equation does **not** represent a linear function?

A. $y = 3x$ **C.** $y = 3x^x$

B. $y = \frac{1}{3}x$ **D.** $y = x - 3$

6. What is the rate of change for the function graphed below?

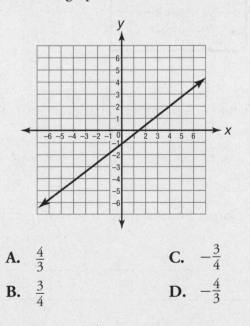

A. $\frac{4}{3}$ **C.** $-\frac{3}{4}$

B. $\frac{3}{4}$ **D.** $-\frac{4}{3}$

7. Which graph does **not** represent a function?

A.

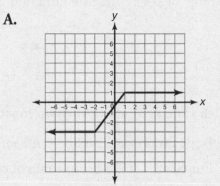

B.

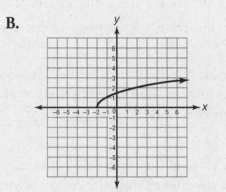

C.

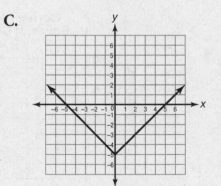

D.

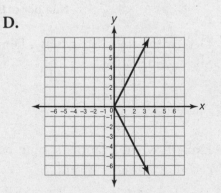

8. For Function 1, any y-value can be determined by using this expression: $\frac{5}{2}x + 3$.

The table below represents Function 2.

x	-1	0	1	2	3
y	-1	3	7	11	15

Which statement about the two functions is true?

A. Both functions have the same rate of change.

B. Function 1 has a greater rate of change than Function 2.

C. Function 2 has a greater rate of change than Function 1.

D. Function 1 has a positive rate of change, and Function 2 has a negative rate of change.

9. Jared is building a tree house. He pays $75 for materials and pays his friend $12 per hour to help him. If Jared spends a total of $129 on building his tree house, for how many hours did his friend work on it? _____

10. The price of buying x boxes of cereal at the Corner Store is shown by the graph below. The advertisement shows the cost of buying x boxes of cereal at Bargain World.

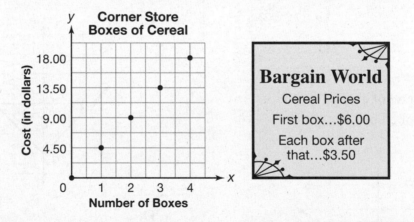

A. Determine the costs when a family buys 1, 2, 3, and 4 boxes of cereal at each store. Show your work.

B. Explain when a family should buy cereal at the Corner Store and when they should buy cereal at Bargain World.

_____

_____

Geometry

Domain 4: Diagnostic Assessment for Lessons 24–32

Lesson 24 Congruence Transformations
8.G.1.a, 8.G.1.b, 8.G.1.c, 8.G.2, 8.G.3

Lesson 25 Dilations
8.G.3, 8.G.4

Lesson 26 Similar Triangles
8.G.4, 8.G.5

Lesson 27 Interior and Exterior Angles of Triangles
8.G.5

Lesson 28 Parallel Lines and Transversals
8.G.5

Lesson 29 The Pythagorean Theorem
8.G.6, 8.G.7

Lesson 30 Distance
8.G.8

Lesson 31 Apply the Pythagorean Theorem
8.G.7, 8.G.8

Lesson 32 Volume
8.G.9

Domain 4: Cumulative Assessment for Lessons 24–32

Domain 4: Diagnostic Assessment for Lessons 24–32

1. Which are the dimensions of a right triangle?

    A. 3 cm, 5 cm, 7 cm

    B. 5 cm, 12 cm, 15 cm

    C. 7 cm, 49 cm, 51 cm

    D. 11 cm, 60 cm, 61 cm

2. A right triangle is formed by squares P, Q, and R. Square P has an area of 25 square meters. Square Q has an area of 144 square meters. The formula for finding the area of a square is $A = s^2$, where s is the length of a side of the square.

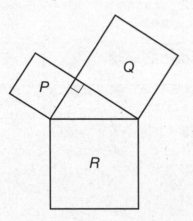

    What is the length of the hypotenuse?

    A. 7 meters

    B. 11 meters

    C. 13 meters

    D. 15 meters

Use the diagram for questions 3 and 4.

Lines r and s are parallel, and line t is a transversal.

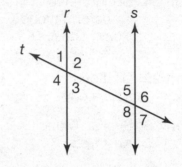

3. If $m\angle 3 = (7x + 9)°$ and $m\angle 5 = (10x - 6)°$, what is the value of x?

    A. $x = 5$

    B. $x = 6$

    C. $x = 7$

    D. $x = 8$

4. If $m\angle 8 = 106°$ and $m\angle 3 = (6x - 4)°$, what is the value of x?

    A. $x = 11$

    B. $x = 12$

    C. $x = 13$

    D. $x = 18$

5. A company wants to lay cable across a lake. To determine the distance across the lake, it took the following measurements and created this diagram.

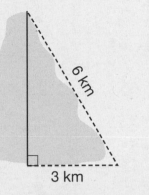

To the nearest tenth of a kilometer, what is the approximate distance across the lake?

A. 2.7 kilometers

B. 3.0 kilometers

C. 5.2 kilometers

D. 6.7 kilometers

6. Deborah bought a spherical exercise ball for her pet hamster. If its volume is 288π cubic inches, what is its radius?

A. 6 inches

B. 12 inches

C. 16 inches

D. 216 inches

7. Two angles of a triangle each measure 38°. What is the measure of the third angle of the triangle?

A. 14°

B. 76°

C. 104°

D. 142°

8. A triangle is graphed on the coordinate grid below.

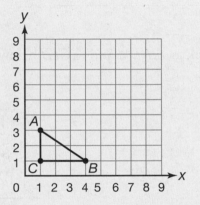

What are the coordinates of the image of this triangle after a dilation, with a center of dilation at the origin and a scale factor of 2?

A. $A'(2, 3)$, $B'(8, 1)$, $C'(2, 1)$

B. $A'(1, 6)$, $B'(4, 2)$, $C'(1, 2)$

C. $A'(2, 6)$, $B'(6, 4)$, $C'(2, 4)$

D. $A'(2, 6)$, $B'(8, 2)$, $C'(2, 2)$

9. What is the length of $\overline{CD}$, in units?

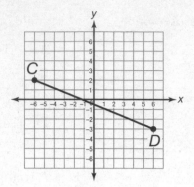

_____

10. Look at $\triangle KLM$ and $\triangle GHI$ below.

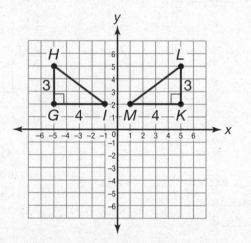

A. Explain how you could prove that $\triangle KLM$ is congruent to $\triangle GHI$.

_____

_____

_____

_____

B. List all the pairs of corresponding congruent sides and corresponding congruent angles for these triangles.

_____

_____

_____

Congruence Transformations

Common Core State Standards:
8.G.1.a, 8.G.1.b, 8.G.1.c, 8.G.2, 8.G.3

Getting the Idea

Congruent figures have the same shape and the same size. A **rigid transformation** is a change in the position of a figure. It does not change the size or the shape of a figure.

Three types of rigid, or congruent, transformations are:

A **translation** is a slide of a figure to a new position.

A **rotation** is a turn of a figure about a point.

A **reflection** is a flip of a figure over a line.

If a figure is translated, rotated, or reflected across a line, the figure and its **image** are congruent. This means that line segments in the figures are the same length, angles have the same measure, and parallel lines remain parallel.

The symbol for congruent is ≅. Congruent figures have the following properties:

The corresponding sides of congruent figures are congruent.

The corresponding angles of congruent figures are congruent.

Example 1

Triangle *FGH* is congruent to triangle *STU*.

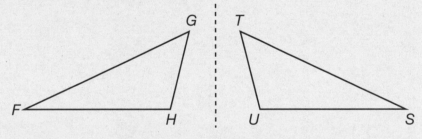

What type of transformation of △*FGH* was used to create △*STU*?

Strategy   **Use the corresponding parts to identify the transformation.**

> Step 1   Look for a set of corresponding angles.

> *F* is the first letter in the name of △*FGH*.

> *S* is the first letter in the name of △*STU*.

> ∠*F* and ∠*S* are corresponding angles.

Step 2   Find other sets of corresponding angles.

∠G and ∠T are corresponding angles.

∠H and ∠U are corresponding angles.

Step 3   Look for a set of corresponding sides.

The first two letters in the name of △FGH are FG.

The first two letters in the name of △STU are ST.

$\overline{FG}$ and $\overline{ST}$ are corresponding sides.

Step 4   Find other sets of corresponding sides.

$\overline{GH}$ and $\overline{TU}$ are corresponding sides.

$\overline{FH}$ and $\overline{SU}$ are corresponding sides.

Step 5   Identify the transformation.

$\overline{FG}$ was flipped over the line to create $\overline{ST}$.

$\overline{GH}$ was flipped over the line to create $\overline{TU}$.

$\overline{FH}$ was flipped over the line to create $\overline{SU}$.

A flip is a reflection.

Solution   **Triangle *FGH* was reflected over a line to create triangle *STU*.**

When a figure is transformed, it creates an image. On a coordinate grid, the vertices of the original figure are labeled with letters, such as *ABCD*. The vertices of the image are labeled with a prime symbol, ('), such as *A′B′C′D′*.

Example 2

Rectangle *EFGH* is translated 5 units right and 6 units down to create rectangle *E'F'G'H'*.

Which pairs of sides are congruent?

Which pairs of sides are parallel?

What are the lengths of the sides?

| **Strategy** | **Use the transformation to identify congruent parts.** |

Step 1   Identify the corresponding sides.

$\overline{EF}$ and $\overline{E'F'}$ are corresponding sides.

$\overline{FG}$ and $\overline{F'G'}$ are corresponding sides.

$\overline{EH}$ and $\overline{E'H'}$ are corresponding sides.

$\overline{HG}$ and $\overline{H'G'}$ are corresponding sides.

Step 2   Identify the congruent sides.

The corresponding sides are congruent.

The opposite sides of a rectangle are congruent.

$\overline{EF} \cong \overline{E'F'} \cong \overline{HG} \cong \overline{H'G'}$.

$\overline{EH} \cong \overline{E'H'} \cong \overline{FG} \cong \overline{F'G'}$.

Step 3   Identify the parallel sides.

The opposite sides of a rectangle are parallel.

$\overline{EF}$ is parallel to $\overline{HG}$, so $\overline{E'F'}$ is parallel to $\overline{H'G'}$.

$\overline{EH}$ is parallel to $\overline{FG}$, so $\overline{E'H'}$ is parallel to $\overline{F'G'}$.

Step 4   Find the lengths of the sides.

$EF = E'F' = HG = H'G' = 3$ units

$EH = E'H' = FG = F'G' = 4$ units

Solution $\overline{EF} \cong \overline{E'F'} \cong \overline{HG} \cong \overline{H'G'}$ and $\overline{EH} \cong \overline{E'H'} \cong \overline{FG} \cong \overline{F'G'}$.
$\overline{EF}$ is parallel to $\overline{HG}$, $\overline{E'F'}$ is parallel to $\overline{H'G'}$,
$\overline{EH}$ is parallel to $\overline{FG}$, and $\overline{E'H'}$ is parallel to $\overline{F'G'}$.
$EF = E'F' = HG = H'G' = 3$ units
$EH = E'H' = FG = F'G' = 4$ units

Example 3

Figure *ABCDE* is rotated 90° counterclockwise about the origin.

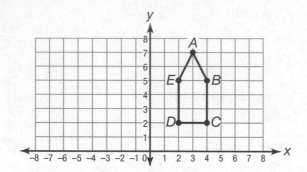

Draw the image of the transformation after it has been rotated.

| **Strategy** | **Think about the definition of rotation.** |

Step 1 Describe the rotation.

A 90° counterclockwise turn is $\frac{1}{4}$ turn to the left.

The point of rotation is the origin.

Step 2 Imagine turning the figure about the origin, (0, 0).

The vertex at *A*(3, 7) moves to *A′*(−7, 3).

The vertex at *B*(4, 5) moves to *B′*(−5, 4).

The vertex at *C*(4, 2) moves to *C′*(−2, 4).

The vertex at *D*(2, 2) moves to *D′*(−2, 2).

The vertex at *E*(2, 5) moves to *E′*(−5, 2).

The coordinates of an original vertex and its image follow a pattern.

The *x*-coordinate of the image is the opposite *y*-coordinate of the original.

The *y*-coordinate of the image is the *x*-coordinate of the original.

Step 3 Plot the vertices and draw the image.

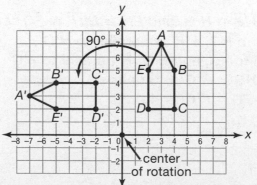

| **Solution** | **Figure *A′B′C′D′E′*, the image of Figure *ABCDE* after a 90° counterclockwise rotation, is shown in Step 3.** |

Coached Example

Trapezoid *QRST* is rotated 180° counterclockwise about the origin to create trapezoid *Q′R′S′T′*.

Which sides of the trapezoids are congruent? Which sides are parallel?

Trapezoid *Q′R′S′T′* is a rotation of trapezoid *QRST*.

So, the figures are the same shape and the same size, or _____.

How many units long is $\overline{QR}$? _____

How many units long is $\overline{Q′R′}$? _____

How many units long is $\overline{ST}$? _____

How many units long is $\overline{S′T′}$? _____

How many units long is $\overline{TQ}$? _____

How many units long is $\overline{T′Q′}$? _____

The corresponding sides of congruent figures are _____.

$\overline{QR}$ corresponds to _____, so $\overline{QR} \cong$ _____.

$\overline{RS}$ corresponds to _____, so $\overline{RS} \cong$ _____.

$\overline{ST}$ corresponds to _____, so $\overline{ST} \cong$ _____.

$\overline{TQ}$ corresponds to _____, so $\overline{TQ} \cong$ _____.

Which sides of trapezoid *QRST* are parallel? _____

So, which sides of trapezoid *Q′R′S′T′* are parallel? _____

$\overline{QR} \cong$ _____, $\overline{RS} \cong$ _____, $\overline{ST} \cong$ _____, and $\overline{TQ} \cong$ _____.

In trapezoid *QRST*, _____ is parallel to _____.

In trapezoid *Q′R′S′T′*, _____ is parallel to _____.

Lesson Practice

Choose the correct answer.

Use the transformation below for questions 1 and 2.

Triangle *ABC* is rotated 90° clockwise to create Triangle *EDC*.

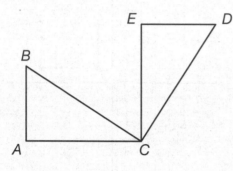

Use the coordinate plane below for questions 3–5.

Parallelogram *QRST* is reflected over the *y*-axis to create parallelogram *Q′R′S′T′*.

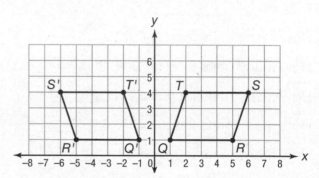

1. Which angle is congruent to ∠*B*?

   A. ∠*BCE*

   B. ∠*ECD*

   C. ∠*E*

   D. ∠*D*

3. Which angle is congruent to ∠*S*?

   A. ∠*Q′*

   B. ∠*R′*

   C. ∠*S′*

   D. ∠*T′*

2. Which side is congruent to $\overline{AC}$?

   A. $\overline{BC}$

   B. $\overline{EC}$

   C. $\overline{ED}$

   D. $\overline{DC}$

4. Which side is congruent to $\overline{QT}$?

   A. $\overline{Q′T′}$

   B. $\overline{ST}$

   C. $\overline{Q′R′}$

   D. $\overline{QR}$

5. $\overline{QR}$ is parallel to $\overline{TS}$. Which side is parallel to $\overline{Q′R′}$?

   A. $\overline{S′R′}$

   B. $\overline{S′T′}$

   C. $\overline{Q′T′}$

   D. $\overline{RS}$

6. △*FGH* is reflected over the *x*-axis.

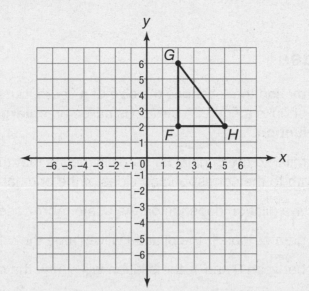

A. Draw the image of the figure after it has been reflected over the *x*-axis. Label the image △*F'G'H'*.

B. Name the congruent sides and the congruent angles.

_____

_____

_____

Dilations

Common Core State Standards:
8.G.3, 8.G.4

Getting the Idea

A **dilation** is a transformation that changes the size of a figure, but not its shape. A dilation stretches or shrinks a figure. A stretch is called an **enlargement**. A shrink is called a **reduction**.

The dilated image is similar to the original figure. This means that the image has sides that are proportional in length to the corresponding sides of the original figure.

The size of a figure after a dilation depends on the **scale factor** used to create it.

• If the scale factor is greater than 1, the dilation will enlarge the original figure.

• If the scale factor is between 0 and 1, the dilation will shrink the original figure.

When working with a dilation, you need to know the scale factor and where the center of dilation is. If the center of dilation is at the origin, you can multiply the coordinates of each vertex by the scale factor to find the vertices of the dilated figure.

Example 1

Rectangle *JKLM* was dilated to form rectangle *J'K'L'M'*.

The center of dilation was at the origin. What scale factor was used?

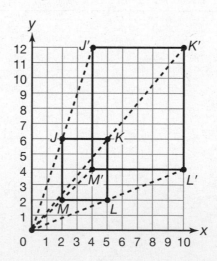

Strategy   **Look at the side lengths of the figure and those of its image.**

Step 1	Decide if the dilation is an enlargement or a reduction.

Rectangle $J'K'L'M'$ is bigger than rectangle $JKLM$.
So, it is an enlargement.

This means that the scale factor is greater than 1.

Step 2	Look at the side lengths of one set of corresponding sides.

$\overline{JK}$ and $\overline{J'K'}$ are horizontal, so you can determine their lengths by counting units.

$J'K' = 6$ units and $JK = 3$ units.

$J'K'$ is 2 times as long as JK. The scale factor is 2.

Step 3	Use the vertices to check your answer.

If you multiply the coordinates of the vertices of the original figure by 2, you should get the coordinates of the vertices of the dilated image.

$J(2, 6)$ → $(2 \times 2, 6 \times 2)$ → $J'(4, 12)$

$K(5, 6)$ → $(5 \times 2, 6 \times 2)$ → $K'(10, 12)$

$L(5, 2)$ → $(5 \times 2, 2 \times 2)$ → $L'(10, 4)$

$M(2, 2)$ → $(2 \times 2, 2 \times 2)$ → $M'(4, 4)$ ✓

Solution   **The scale factor for the dilation was 2.**

Example 2

Dilate $\triangle FGH$ with the center of dilation at the origin and a scale factor of $\frac{1}{3}$.

Draw the dilated image and identify the coordinates of its vertices.

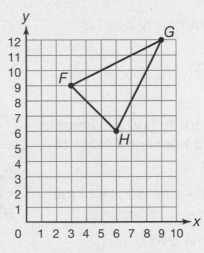

Strategy   **Use the scale factor to identify the coordinates of the vertices of the image.**

Step 1   Multiply the coordinates of each vertex of $\triangle FGH$ by the scale factor, $\frac{1}{3}$.

$$F(3, 9) \rightarrow \left(3 \times \tfrac{1}{3}, 9 \times \tfrac{1}{3}\right) \rightarrow F'(1, 3)$$

$$G(9, 12) \rightarrow \left(9 \times \tfrac{1}{3}, 12 \times \tfrac{1}{3}\right) \rightarrow G'(3, 4)$$

$$H(6, 6) \rightarrow \left(6 \times \tfrac{1}{3}, 6 \times \tfrac{1}{3}\right) \rightarrow H'(2, 2)$$

Step 2   Plot and connect the vertices of the image, $\triangle F'G'H'$.

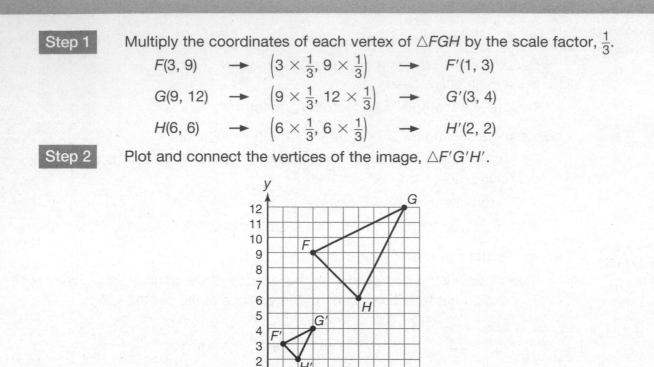

Solution   **The vertices of the dilated image are $F'(1, 3)$, $G'(3, 4)$, and $H'(2, 2)$. Its graph is shown in Step 2.**

Remember, the center of dilation is not always at the origin.

Example 3

Dilate $\triangle ABC$ with the center of dilation at point B and a scale factor of 3.

Draw the dilated image and identify the coordinates of its vertices.

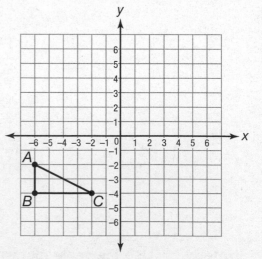

Strategy   **Use the scale factor to find the lengths of the vertical and horizontal sides of the dilated image.**

Step 1   Think about how the side lengths of the figure change with a scale factor of 3.

Each side of the dilated image will be 3 times as long as its corresponding side on $\triangle ABC$.

Step 2   Find the length of side $\overline{A'B}$.

The length of $\overline{AB}$ is 2 units.

$\overline{A'B} = 2 \times 3 = 6$ units

The center of dilation is at point B, so count up 6 units from point B.

Plot point A' there, at $(-6, 2)$.

Note: Do not write a prime symbol after the letter B because the location of point B is the same for the original triangle and its image.

Step 3   Find the length of side $\overline{BC'}$.

The length of $\overline{BC}$ is 4 units.

$\overline{BC'} = 4 \times 3 = 12$ units

Count 12 units to the right of point B.

Plot point C' there, at $(6, -4)$.

Connect the vertices.

Solution   The vertices of the dilated image are $A'(-6, 2)$, $B(-6, -4)$, and $C'(6, -4)$. Its graph is shown in Step 2.

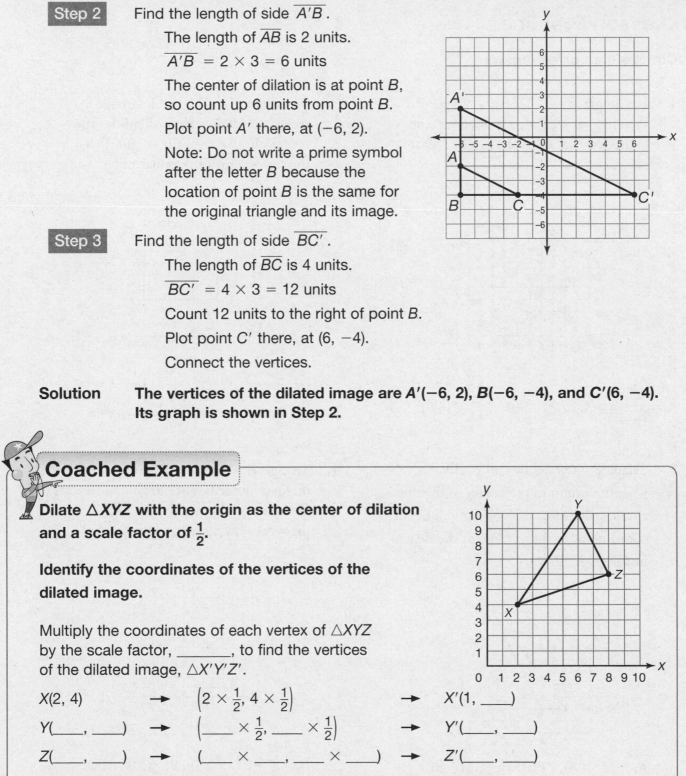

Coached Example

Dilate $\triangle XYZ$ with the origin as the center of dilation and a scale factor of $\frac{1}{2}$.

Identify the coordinates of the vertices of the dilated image.

Multiply the coordinates of each vertex of $\triangle XYZ$ by the scale factor, _____, to find the vertices of the dilated image, $\triangle X'Y'Z'$.

$X(2, 4)$ → $\left(2 \times \frac{1}{2}, 4 \times \frac{1}{2}\right)$ → $X'(1,$ ___$)$

$Y($___, ___$)$ → $\left(\_\_ \times \frac{1}{2}, \_\_ \times \frac{1}{2}\right)$ → $Y'($___, ___$)$

$Z($___, ___$)$ → $(\_\_ \times \_\_, \_\_ \times \_\_)$ → $Z'($___, ___$)$

Plot the vertices of $\triangle X'Y'Z'$ on the grid above. Then connect them.

The dilated image, with vertices $X'(1,$ ___$)$, $Y'($___, ___$)$, and $Z'($___, ___$)$, is graphed above.

193

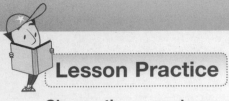

Lesson Practice

Choose the correct answer.

1. Rectangle $W'X'Y'Z'$ is the image of rectangle $WXYZ$ after a dilation. The center of dilation is the origin. What is the scale factor of the dilation?

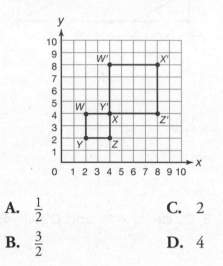

A. $\frac{1}{2}$ C. 2

B. $\frac{3}{2}$ D. 4

2. Triangle NOP below will be dilated with the origin as the center of dilation and a scale factor of $\frac{1}{2}$. What will be the coordinates of the vertices of the dilated image, $\triangle N'O'P'$?

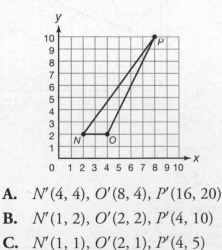

A. $N'(4, 4), O'(8, 4), P'(16, 20)$

B. $N'(1, 2), O'(2, 2), P'(4, 10)$

C. $N'(1, 1), O'(2, 1), P'(4, 5)$

D. $N'(1, 1), O'(1, 1), P'(3, 5)$

3. Rectangle $A'B'C'D'$ is the image of rectangle $ABCD$ after a dilation. The center of dilation is the origin. What is the scale factor of the dilation?

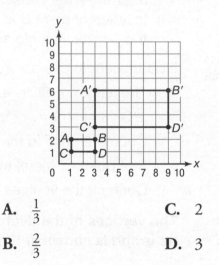

A. $\frac{1}{3}$ C. 2

B. $\frac{2}{3}$ D. 3

4. Triangle RST below will be dilated with the origin as the center of dilation and a scale factor of 4. What will be the coordinates of the vertices of the dilated image, $\triangle R'S'T'$?

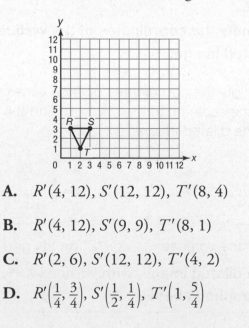

A. $R'(4, 12), S'(12, 12), T'(8, 4)$

B. $R'(4, 12), S'(9, 9), T'(8, 1)$

C. $R'(2, 6), S'(12, 12), T'(4, 2)$

D. $R'\left(\frac{1}{4}, \frac{3}{4}\right), S'\left(\frac{1}{2}, \frac{1}{4}\right), T'\left(1, \frac{5}{4}\right)$

5. Quadrilateral $J'KL'M'$ is the image of quadrilateral $JKLM$ after a dilation. Point K is the center of dilation. What is the scale factor of the dilation?

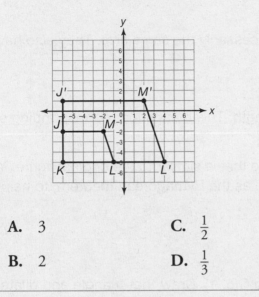

A. 3

B. 2

C. $\dfrac{1}{2}$

D. $\dfrac{1}{3}$

6. Triangle ABC is dilated with a scale factor of 4 and the center of dilation is the origin. If the coordinates of the vertices of $\triangle ABC$ are $A(4, 8)$, $B(12, 8)$, and $C(16, 20)$, what are the coordinates of the vertices of the image after the dilation?

A. $A'(16, 32)$, $B'(48, 32)$, $C'(64, 80)$

B. $A'(16, 32)$, $B'(48, 8)$, $C'(48, 80)$

C. $A'(8, 12)$, $B'(16, 12)$, $C'(20, 24)$

D. $A'(1, 2)$, $B'(3, 2)$, $C'(4, 5)$

7. Triangle ABC has vertices $A(1, 6)$, $B(5, 4)$, and $C(3, 1)$. Triangle ABC will be dilated by a scale factor of 3, with the center of dilation at the origin.

A. Find the coordinates of the vertices of the dilated image, $\triangle A'B'C'$. Show your work.

_____

_____

_____

B. On the coordinate grid, graph $\triangle ABC$ and its image, $\triangle A'B'C'$.

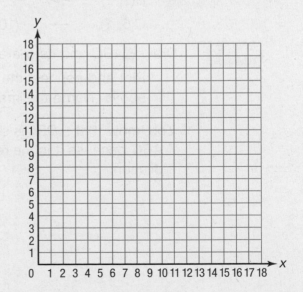

Similar Triangles

Common Core State Standards:
8.G.4, 8.G.5

Getting the Idea

Similar triangles have the same shape but not necessarily the same size. They also have the following properties:

- Their corresponding angles are congruent.

- Their corresponding sides are proportional in length. This means that corresponding sides in similar triangles have the same ratio.

You already know that a dilation results in an image that is similar to the original figure. You can use that fact, and some other knowledge such as the Pythagorean theorem, to help you understand why two triangles are similar.

Example 1

Right triangle ABC has vertices A(1, 4), B(1, 1) and C(5, 1). Draw that triangle and dilate it with the origin as the center of dilation and a scale factor of 2. Name the image △DEF. Compare the side lengths and angle measures of the triangles. What do they tell you about the triangles?

Strategy   **Graph the result of the dilation. Then compare the corresponding parts of the triangles.**

Step 1   Multiply the coordinates of the vertices of △ABC by the scale factor, 2.

$A(1, 4) \longrightarrow (2, 8)$

$B(1, 1) \longrightarrow (2, 2)$

$C(5, 1) \longrightarrow (10, 2)$

Step 2   Draw △ABC and its dilated image.

Plot and connect the coordinates above. Name the image △DEF.

You know that △DEF is similar to △ABC because it is the result of a dilation.

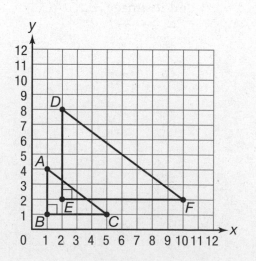

Step 3	Compare the vertical and horizontal side lengths of the two triangles.

You can count units to determine the horizontal and vertical side lengths of $\triangle ABC$ and $\triangle DEF$.

AB = 3 units, and its corresponding side, $\overline{DE}$, measures 6 units.

BC = 4 units, and its corresponding side, $\overline{EF}$, measures 8 units.

Step 4	Compare the other side lengths.

Since you know the lengths of two legs of both triangles, you can use the Pythagorean theorem to find the lengths of the hypotenuses.

Hypotenuse of $\triangle ABC$:

$$a^2 + b^2 = c^2$$
$$3^2 + 4^2 = c^2$$
$$25 = c^2$$
$$5 = c$$

Hypotenuse of $\triangle DEF$:

$$a^2 + b^2 = c^2$$
$$6^2 + 8^2 = c^2$$
$$100 = c^2$$
$$10 = c$$

So, AC = 5 units, and its corresponding side, $\overline{DF}$, measures 10 units.

Step 5	Compare the corresponding sides.

$$\frac{AB}{DE} = \frac{BC}{EF} = \frac{AC}{DF}$$

$$\frac{3}{6} = \frac{4}{8} = \frac{5}{10}$$

The ratio of each pair of corresponding sides is equivalent to $\frac{1}{2}$.

So, corresponding sides have proportional lengths.

Step 6	Compare the known angle measures.

Angle E measures $90°$ because it is a right angle, just like its corresponding angle, $\angle B$.

Step 7	Use a protractor to compare the other angle measures.

$\angle A$ is congruent to $\angle D$, and $\angle C$ is congruent to $\angle F$.

This means that all corresponding angles of the two triangles are congruent.

Solution   **The two triangles are similar because $\triangle DEF$ is the dilated image of $\triangle ABC$. Their corresponding sides have proportional lengths, and their corresponding angles are congruent.**

To prove that two triangles are similar, you do not need to show that all three pairs of angles are congruent and all three pairs of sides have proportional lengths. That would take a long time. Example 2 shows a shorter way.

Example 2

Right triangle *NOP* has vertices *N*(2, 10), *O*(14, 10) and *P*(14, 15). Draw that triangle and determine the coordinates of its image after a dilation with the center of dilation at the origin and a scale factor of $\frac{1}{2}$. Plot those coordinates and draw two sides of the smaller triangle, named △*QRS*. Can you prove that the two triangles are similar?

Strategy **Graph the original triangle and the vertices of its dilation. Then compare corresponding parts of the triangles.**

Step 1 Multiply the coordinates of the vertices of △*NOP* by the scale factor, $\frac{1}{2}$.

N(2, 10) → (1, 5)

O(14, 10) → (7, 5)

P(14, 15) → (7, 7.5)

Step 2 Draw △*NOP* and plot the coordinates of its dilated image.

Plot the coordinates and connect the points for *N*, *O*, and *P*.

Then, name and connect points *Q* and *R* and points *S* and *R* to form two sides of the dilated image.

Step 3 Compare the sides and angles.

Count units to determine if the side lengths are proportional.

$\frac{NO}{QR} = \frac{PO}{SR}$

$\frac{12}{6} = \frac{5}{2.5}$

Both ratios above are equivalent to 2.

So, two pairs of corresponding sides have proportional lengths.

The angles between them are congruent because when right triangle *NOP* is dilated, its image is a right triangle.

So, $m\angle O = m\angle R = 90°$.

Step 4   Decide if it is possible for $\overline{QS}$ to have a length that is not proportional to $\overline{NP}$.

There is only one possible length for the third side of $\triangle QRS$.

That length will be proportional to NP.

You can prove this using the Pythagorean theorem.

Hypotenuse of $\triangle NOP$:

$$a^2 + b^2 = c^2$$
$$12^2 + 5^2 = c^2$$
$$169 = c^2$$
$$13 = c$$

Hypotenuse of $\triangle QRS$:

$$a^2 + b^2 = c^2$$
$$6^2 + 2.5^2 = c^2$$
$$42.25 = c^2$$
$$6.5 = c$$

$NP = 13$ units, and its corresponding side, QS, measures 6.5 units.

$$\frac{NP}{QS} = \frac{13}{6.5} = 2$$

This is the same ratio as the other pairs of corresponding sides.

Solution   **When two pairs of corresponding sides have proportional lengths and the corresponding angles between them are congruent, triangles are similar.**

Note: Example 2 shows the **SAS (Side-Angle-Side) similarity theorem.**

Another way to prove that two triangles are similar is to prove that two pairs of corresponding angles have the same measure. Remember that the sum of the measures of the angles of a triangle is 180°.

Example 3

Triangle *KLM* is the result of a dilation of △*GHI*.

Compare corresponding angles to prove that these two triangles are similar. How many pairs of corresponding angles do you need to compare in order to prove that the two triangles are similar?

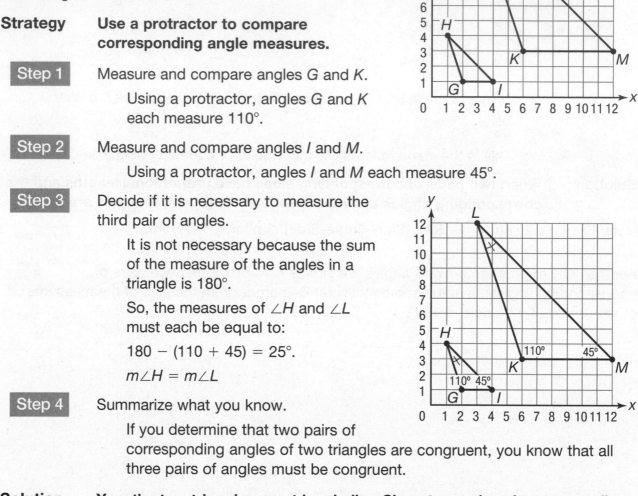

Strategy   **Use a protractor to compare corresponding angle measures.**

Step 1   Measure and compare angles *G* and *K*.

     Using a protractor, angles *G* and *K* each measure 110°.

Step 2   Measure and compare angles *I* and *M*.

     Using a protractor, angles *I* and *M* each measure 45°.

Step 3   Decide if it is necessary to measure the third pair of angles.

     It is not necessary because the sum of the measure of the angles in a triangle is 180°.

     So, the measures of ∠*H* and ∠*L* must each be equal to:

     180 − (110 + 45) = 25°.

     $m\angle H = m\angle L$

Step 4   Summarize what you know.

     If you determine that two pairs of corresponding angles of two triangles are congruent, you know that all three pairs of angles must be congruent.

Solution   **Yes, the two triangles must be similar. Since two pairs of corresponding angles are congruent, all three pairs of angles must be congruent.**

Note: Example 3 shows the **AA (Angle-Angle) similarity theorem.**

Now that you know how to use dilations and coordinate grids to prove that two triangles are similar, you can apply that knowledge to find missing side lengths or missing angle measures in two similar triangles.

Example 4

What is the value of x in the diagram below?

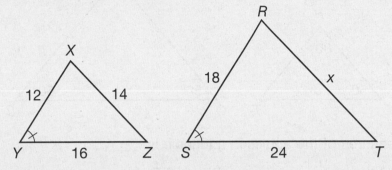

Strategy **Determine if the two triangles are similar. If so, set up and solve a proportion to find the value of x.**

Step 1 Decide if the two triangles are similar.

If two pairs of corresponding sides have proportional lengths and the angles between them are congruent, then the triangles are similar.

The diagram gives lengths for corresponding sides XY and RS and for corresponding sides YZ and ST.

It also shows that the corresponding angles Y and S between these sides are congruent.

$$\frac{XY}{RS} = \frac{YZ}{ST}$$

$$\frac{12}{18} = \frac{16}{24}$$

Both ratios are equivalent to $\frac{2}{3}$. So, those sides have proportional lengths.

Triangle XYZ is similar to Triangle RST.

Step 2 Set up a proportion that could be used to find the value of x.

$\frac{XZ}{RT} = \frac{14}{x}$, so:

$$\frac{14}{x} = \frac{2}{3}$$

Step 3 Cross-multiply to solve for x.

$$\frac{14}{x} = \frac{2}{3}$$

$$14 \cdot 3 = 2x$$

$$42 = 2x \qquad \text{Divide both sides by 2.}$$

$$21 = x$$

Solution **The value of x is 21 units.**

Coached Example

Are the two triangles shown below similar? Explain how you know.

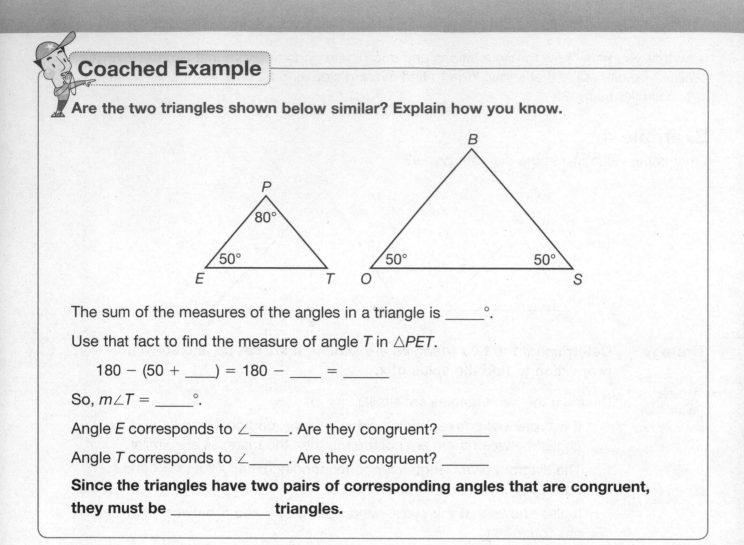

The sum of the measures of the angles in a triangle is _____°.

Use that fact to find the measure of angle *T* in △*PET*.

    180 − (50 + ____) = 180 − ____ = _____

So, *m*∠*T* = _____°.

Angle *E* corresponds to ∠_____. Are they congruent? _____

Angle *T* corresponds to ∠_____. Are they congruent? _____

Since the triangles have two pairs of corresponding angles that are congruent, they must be _____ triangles.

Lesson Practice

Choose the correct answer.

Use the diagram for questions 1 and 2.

Triangle *RST* is the result of a dilation of △*NPQ* with the center of dilation at the origin and a scale factor of $\frac{1}{3}$.

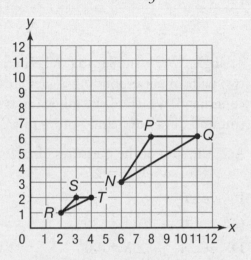

1. Which of the following must be true?

    A. ∠*N* is congruent to ∠*R*.

    B. ∠*N* is congruent to ∠*S*.

    C. ∠*P* is congruent to ∠*R*.

    D. ∠*P* is congruent to ∠*T*.

2. Which proportion must be correct?

    A. $\dfrac{NP}{RS} = \dfrac{PQ}{RT}$

    B. $\dfrac{NP}{RS} = \dfrac{PQ}{ST}$

    C. $\dfrac{NP}{RT} = \dfrac{PQ}{ST}$

    D. $\dfrac{NP}{RT} = \dfrac{NQ}{RS}$

Use the diagram for questions 3 and 4.

Triangle *XYZ* is the result of a dilation of △*JKL* with the center of dilation at the origin and a scale factor of 4.

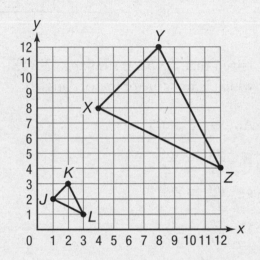

3. Which names a pair of corresponding, congruent angles?

    A. ∠*L* and ∠*X*   **C.** ∠*K* and ∠*X*

    B. ∠*L* and ∠*Y*   **D.** ∠*K* and ∠*Y*

4. Which is **not** true of the triangles in the diagram?

    A. △*XYZ* is similar to △*JKL* because a dilated image is similar to the original figure.

    B. The ratio $\dfrac{JK}{XY}$ is equivalent to the ratio $\dfrac{KL}{YZ}$.

    C. $m\angle J = m\angle X$

    D. $JL = XZ$

Use the diagram for questions 5 and 6.

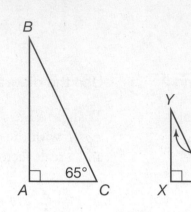

5. Given that △ABC is similar to △XYZ, which is **not** true of these triangles?

    A. $m\angle Z = 65°$

    B. $\dfrac{BC}{YZ} = \dfrac{BA}{XZ}$

    C. $\dfrac{BC}{YZ} = \dfrac{BA}{YX}$

    D. $\dfrac{BC}{YZ} = \dfrac{AC}{XZ}$

6. Triangle ABC is similar to △XYZ because two pairs of corresponding angles are congruent. Which of the following is **not** true and does **not** help to prove this?

    A. The sum of the measures of △ABC is 180°, so $m\angle B = 25°$.

    B. $m\angle B = m\angle Y = 25°$

    C. $m\angle C = m\angle Y = 25°$

    D. $m\angle A = m\angle X = 90°$

7. Look at △HJK and △DEF.

    A. Explain how you could prove that △HJK is similar to △DEF.

    _____

    _____

    _____

    B. List all the pairs of corresponding, congruent angles. Then list all the ratios of corresponding, proportional side lengths for these triangles.

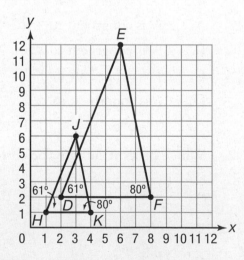

    _____

    _____

    _____

Common Core State Standard:
8.G.5

Interior and Exterior Angles of Triangles

Getting the Idea

A triangle is a polygon with 3 angles and 3 sides. The sum of the measures of the angles of a triangle is 180°. The diagram below shows this.

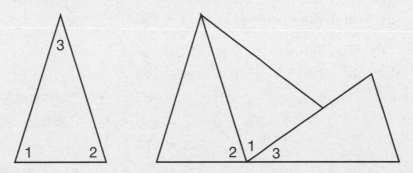

Look at the triangle on the left. If you make three copies of the triangle and rotate them so that the 3 angles are aligned, you can see that the angles form a **straight angle**, which has a measure of 180°.

Example 1

What is the missing angle measure in the triangle below?

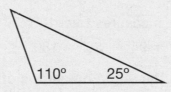

Strategy	Use the sum of the angle measures of a triangle.
Step 1	Find the sum of the two known angle measures.
	110° + 25° = 135°
Step 2	Subtract the sum from the total angle measures of a triangle.
	180° − 135° = 45°
Solution	**The missing angle measure is 45°.**

Example 2

The angles of a triangle measure $5x + 8$ degrees, $9x + 2$ degrees, and $15x - 4$ degrees.

What is the value of x? What are the measures of the angles of the triangle?

Strategy   **Use the sum of the angle measures of a triangle.**

Step 1   Write an equation.

Write an equation that shows the sum of the angle measures of a triangle equals 180°.

$$(5x + 8) + (9x + 2) + (15x - 4) = 180$$

Step 2   Solve the equation.

$$(5x + 8) + (9x + 2) + (15x - 4) = 180$$

$$29x + 6 = 180 \qquad \text{Combine like terms.}$$

$$29x + 6 - 6 = 180 - 6 \qquad \text{Subtract 6 from both sides.}$$

$$29x = 174$$

$$\frac{29x}{29} = \frac{174}{29} \qquad \text{Divide both sides by 29.}$$

$$x = 6$$

Step 3   Substitute 6 for x in each expression for the angle measures.

$$5x + 8 = (5 \times 6) + 8 = 30 + 8 = 38°$$

$$9x + 2 = (9 \times 6) + 2 = 54 + 2 = 56°$$

$$15x - 4 = (15 \times 6) - 4 = 90 - 4 = 86°$$

Step 4   Check that the angle measures total 180°.

$$38° + 56° + 86° = 94° + 86° = 180° \qquad ✓$$

Solution   **The value of x is 6. The angle measures of the triangle are 38°, 56°, and 86°.**

An **interior angle** of a polygon is an angle that is on the inside of the polygon and has its vertex formed by two sides of the polygon. An **exterior angle** of a polygon is an angle formed by a side of the polygon and an extension of an adjacent side.

In the diagram below, ∠1 is an interior angle of the triangle and ∠2 is an exterior angle of the triangle.

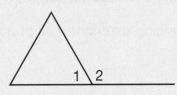

An exterior angle and its corresponding interior angle form a straight angle. So, an exterior angle and the corresponding interior angle are **supplementary**, or have a sum of 180°.

Example 3

An exterior angle of a triangle measures 128°, as shown in the diagram.

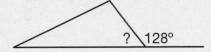

What is the measure of the corresponding interior angle of the triangle?

Strategy	**Use the fact that an exterior angle and its corresponding interior angle are supplementary.**
Step 1	Understand the meaning of supplementary angles.
	Supplementary angles have a sum of 180°.
Step 2	Subtract the exterior angle measure from 180°.
	180° − 128° = 52°
Solution	**The measure of the corresponding interior angle is 52°.**

Coached Example

An angle of a triangle has a measure of 3x + 1 degrees. The corresponding exterior angle has a measure of 5x − 13 degrees. What is the value of x? What are the measures of the interior angle and the exterior angle?

An interior angle and the corresponding exterior angle of a triangle are _____.

Supplementary angles add up to _____.

Write an equation that shows the sum of the measures of the interior and exterior angles equals _____.

_____

Solve the equation for x.

The value of x is _____.

To find the measure of the interior angle, substitute the value of x into _____.
Solve.

_____

To find the measure of the exterior angle, substitute the value of x into _____.
Solve.

_____

Check that the sum of the angle measures is _____.

_____

The value of x is _____. The interior angle measures _____°.

The corresponding exterior angle measures _____°.

Lesson Practice

Choose the correct answer.

1. One angle of a triangle measures 34°. A second angle measures 81°. What is the measure of the third angle of the triangle?

   A. 65°

   B. 99°

   C. 115°

   D. 146°

2. The angle measures of a triangle are $2x + 5$ degrees, $6x - 5$ degrees, and $7x$ degrees. What is the value of x?

   A. $x = 10$

   B. $x = 12$

   C. $x = 15$

   D. $x = 29$

3. An exterior angle of a triangle measures 119°. What is the measure of the corresponding interior angle?

   A. 29°

   B. 61°

   C. 69°

   D. 71°

4. The measure of an interior angle of a triangle is $4n + 5$ degrees. The measure of the corresponding exterior angle is $15n + 4$ degrees. What is the value of n?

   A. $n = 1$

   B. $n = 9$

   C. $n = 10$

   D. $n = 12$

5. The angle measures of a triangle are $2y + 3$ degrees, $9y - 7$ degrees, and $7y + 4$ degrees. Which of the following is **not** one of the angle measures of the triangle?

   A. 23°

   B. 74°

   C. 83°

   D. 97°

6. A triangle has vertices A, B, and C. The measure of $\angle A$ is $2x$ degrees. The measure of $\angle B$ is twice the measure of $\angle A$. The measure of $\angle C$ is three times the measure of $\angle A$. What is the value of x?

   A. $x = 12$

   B. $x = 15$

   C. $x = 24$

   D. $x = 30$

7. The measure of one angle of a right triangle is 74°.

   A. What are the measures of the other two angles in the triangle?
   Explain your answer.

   _____

   _____

   _____

   B. What are the measures of each of the corresponding exterior angles?
   Explain your answer.

   _____

   _____

   _____

Parallel Lines and Transversals

Common Core State Standard:
8.G.5

Getting the Idea

Parallel lines lie in the same plane and never intersect. They are always the same distance apart. A **transversal** is a line that intersects two or more lines.

When a transversal intersects parallel lines, some special angle pairs are formed.

In the diagram below, line j is parallel to line k and line t is a transversal.

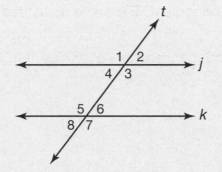

Corresponding angles are angles that lie on the same side of a transversal and on the same side of each parallel line. One pair of corresponding angles is $\angle 1$ and $\angle 5$.

Corresponding angles are congruent.

$\angle 1 \cong \angle 5$ $\qquad$ $\angle 3 \cong \angle 7$

$\angle 2 \cong \angle 6$ $\qquad$ $\angle 4 \cong \angle 8$

Interior angles lie between lines j and k.

The interior angles are $\angle 3$, $\angle 4$, $\angle 5$, and $\angle 6$.

Alternate interior angles lie inside parallel lines and on opposite sides of the transversal. Alternate interior angles are congruent.

$\angle 4 \cong \angle 6$

$\angle 3 \cong \angle 5$

Exterior angles lie outside lines j and k.

The exterior angles are $\angle 1$, $\angle 2$, $\angle 7$, and $\angle 8$.

Alternate exterior angles lie outside parallel lines and on opposite sides of the transversal. Alternate exterior angles are congruent.

$\angle 1 \cong \angle 7$

$\angle 2 \cong \angle 8$

You can arrange congruent triangles to understand why the special angle pairs are congruent.

In the diagram below, line j is parallel to line k and line t is a transversal.

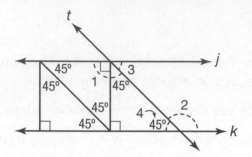

The three triangles shown are congruent. The angle measures of the triangles are 45°, 45°, and 90°.

You can use the diagram to find the measures of angles 1–4.

$m\angle 1 = 90° + 45° = 135°$

$m\angle 3 = 180° - 135° = 45°$

$m\angle 4 = 45°$

$m\angle 2 = 180° - 45° = 135°$

$\angle 1 \cong \angle 2$, and they are alternate interior angles.

$\angle 3 \cong \angle 4$, and they are alternate interior angles.

You can use similar reasoning and supplementary angle relationships to show that the alternate exterior angles are congruent and that the corresponding angles are congruent.

Example 1

Lines c and d are parallel, and line t is a transversal.

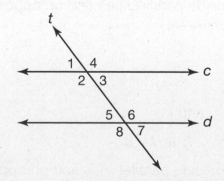

If $m\angle 1 = 75°$, what is $m\angle 8$?

Strategy Use the properties of angle pairs formed by parallel lines and a transversal.

Step 1	Find the corresponding angle to ∠1.
	∠1 and ∠5 are corresponding angles.

Step 2	Find the measure of ∠5.
	∠1 ≅ ∠5, so $m∠5 = m∠1 = 75°$.

Step 3	Find the relationship between ∠5 and ∠8.
	∠5 and ∠8 are supplementary angles.

Step 4	Find the measure of ∠8.
	The sum of the measures of supplementary angles is 180°.

$$m∠5 + m∠8 = 180$$
$$75 + m∠8 = 180$$
$$m∠8 = 180 - 75$$
$$m∠8 = 105°$$

Solution   **The measure of ∠8 is 105°.**

Note: There are other ways to solve Example 1. For example, you can use alternate exterior angles 1 and 7 and then the fact that angles 7 and 8 are supplementary to find the measure of ∠8.

Example 2

Lines *r* and *s* are parallel, and line *a* is a transversal. The measure of ∠4 = 140°.

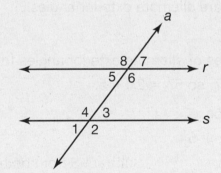

What is the measure of ∠5?

Strategy	**Use the properties of angle pairs formed by parallel lines and a transversal.**

Step 1	Find an angle that is congruent to ∠4.
	∠4 and ∠6 are alternate interior angles.
	∠4 ≅ ∠6

Step 2	Find the measure of ∠6.
	$m∠6 = m∠4 = 140°$

Step 3   Find the relationship between $\angle 5$ and $\angle 6$.

$\angle 5$ and $\angle 6$ are supplementary angles.

Step 4   Use the properties of supplementary angles to find the measure of $\angle 5$.

The sum of the measures of supplementary angles is 180°.

$$m\angle 5 + m\angle 6 = 180$$
$$m\angle 5 + 140 = 180$$
$$m\angle 5 = 180 - 140$$
$$m\angle 5 = 40°$$

Solution   The measure of $\angle 5$ is 40°.

Note: As with Example 1, there are other ways to solve Example 2.

Example 3

Lines a and b are parallel, and line m is a transversal.
If $m\angle 1 = 67°$ and $m\angle 7 = (2x + 5)°$, what is the value of x?

Strategy   Use the properties of angle pairs formed by parallel lines and a transversal.

Step 1   Look for a special angle relationship between $\angle 1$ and $\angle 7$.

$\angle 1$ and $\angle 7$ are alternate exterior angles.

$\angle 1 \cong \angle 7$

Step 2   Use the properties of alternate exterior angles to set up an equation.

$m\angle 1 = m\angle 7$, so $67 = 2x + 5$.

Step 3   Solve the equation.

$$67 = 2x + 5$$
$$62 = 2x \qquad \text{Subtract 5 from both sides.}$$
$$\frac{62}{2} = \frac{2x}{2} \qquad \text{Divide both sides by 2.}$$
$$31 = x$$

Solution   The value of x is 31.

Coached Example

Lines *m* and *n* are parallel, and line *t* is a transversal.

If $m\angle 6 = 130°$, what is $m\angle 3$?

Is $\angle 6$ an interior angle or an exterior angle? _____

$\angle 6$ and $\angle 2$ are _____ _____ angles.

So, $\angle 6$ is _____ to $\angle 2$.

The measure of $\angle 6$ is 130°, so the measure of $\angle 2$ is _____ °.

$\angle 2$ and $\angle 3$ are _____ angles.

The sum of the measures of $\angle 2$ and $\angle 3$ is _____ °.

Find the measure of $\angle 3$.

_____ − _____ = _____

The measure of $\angle 3$ is _____°.

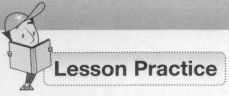

Lesson Practice

Choose the correct answer.

Use the diagram for questions 1–3.

Lines *k* and *l* are parallel, and line *y* is a transversal.

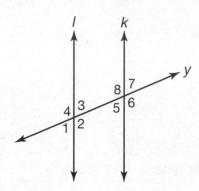

1. Which pair of angles are corresponding angles?

   A. ∠1 and ∠2

   B. ∠2 and ∠5

   C. ∠1 and ∠3

   D. ∠3 and ∠7

2. Which pair of angles are alternate interior angles?

   A. ∠2 and ∠3

   B. ∠2 and ∠8

   C. ∠3 and ∠8

   D. ∠4 and ∠8

3. Which pair of angles are alternate exterior angles?

   A. ∠1 and ∠4   C. ∠4 and ∠6

   B. ∠1 and ∠6   D. ∠5 and ∠7

Use the diagram for questions 4–6.

Lines *p* and *q* are parallel, and line *w* is a transversal.

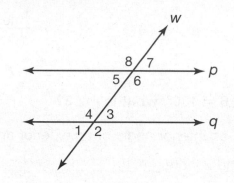

4. Which angle is congruent to ∠6?

   A. ∠1

   B. ∠2

   C. ∠5

   D. ∠7

5. Which angle is **not** congruent to ∠1?

   A. ∠3

   B. ∠5

   C. ∠7

   D. ∠8

6. If ∠2 measures 132°, what is the measure of ∠7?

   A. 42°

   B. 48°

   C. 58°

   D. 132°

7. In the diagram below, lines *a* and *b* are parallel, and line *t* is a transversal.

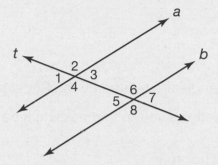

A. If $m\angle 2 = 112°$ and $m\angle 5 = (9x + 5)°$, what is the value of *x*?
Explain your answer and show your work.

_____

_____

_____

_____

B. What are the measures of angles 1 through 8? Explain your answers.

_____

_____

_____

The Pythagorean Theorem

Common Core State Standards:
8.G.6, 8.G.7

Getting the Idea

The side lengths in a right triangle are related by the Pythagorean theorem.

The formula for this theorem allows you to find missing side lengths in right triangles.

The Pythagorean Theorem

In a right triangle, the sum of the squares of the lengths of the legs is equal to the square of the length of the hypotenuse.

$$a^2 + b^2 = c^2$$

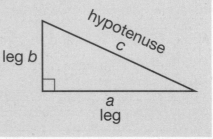

Example 1

Verify the Pythagorean theorem using a model.

Strategy   **Use graph paper to model the sides of a right triangle.**

Step 1   Cut out three squares with sides 3, 4, and 5 units long.

Make a right triangle with the squares.

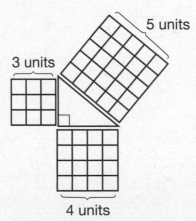

Step 2   Find the area of each square.

The formula for the area of a square is: $A = s^2$, where s is the length of a side.

The area of a square with sides 3 units long is $3^2 = 9$ units^2.

The area of a square with sides 4 units long is $4^2 = 16$ units^2.

The area of a square with sides 5 units long is $5^2 = 25$ units^2.

Step 3	Show how the areas of those squares are related.

$$(3 \text{ units})^2 + (4 \text{ units})^2 = (5 \text{ units})^2$$

$$9 \text{ units}^2 + 16 \text{ units}^2 = 25 \text{ units}^2$$

The sum of the areas of the two smaller squares is equal to the area of the largest square.

So, the sum of the squares of the two smaller side lengths is equal to the square of the largest side length. That is the Pythagorean theorem.

Solution   **Steps 1–3 verify the Pythagorean theorem.**

Below is another proof for this theorem.

A right triangle with legs a and b and hypotenuse c is shown on the right. Its area, in square units, is $\frac{1}{2}ab$.

Now, imagine two congruent squares, each with sides $(a + b)$ units long. See Figures 1 and 2 below.

Figure 1 is divided into 4 congruent right triangles and two shaded squares.
One shaded square has sides a units long, so its area is a^2.
The other shaded square has sides b units long, so its area is b^2.

Figure 2, on the other hand, is divided into 4 congruent right triangles and one shaded square with sides c units long. So, the area of that shaded square is c^2.

Figure 1   **Figure 2**

$(a + b)^2$	$=$	$(a + b)^2$	Both figures have the same area.
$a^2 + b^2 + (4 \times \frac{1}{2} ab)$	$=$	$c^2 + (4 \times \frac{1}{2} ab)$	Subtract $(4 \times \frac{1}{2} ab)$ from both sides.
$- (4 \times \frac{1}{2} ab)$		$- (4 \times \frac{1}{2} ab)$	
$a^2 + b^2$	$=$	c^2	This is the formula for the Pythagorean Theorem

Each large square contains four congruent white triangles. Since the large squares are congruent and the subtracted area of the white triangles is also congruent, then the remaining, shaded areas are congruent as well. The sum of the shaded areas in Figure 1 is equal to the shaded area in Figure 2.

Example 2

What is the length of the third side in this right triangle?

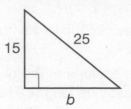

Strategy   **Use the formula for the Pythagorean theorem.**

Step 1   Identify the legs and the hypotenuse.

The legs have lengths of 15 units and b units. The hypotenuse is 25 units long.

Step 2   Use the formula.

Substitute 15 for a and 25 for c into the formula. Solve for b.

$$a^2 + b^2 = c^2$$
$$15^2 + b^2 = 25^2$$
$$225 + b^2 = 625$$
$$b^2 = 400$$
$$b = \sqrt{400}$$
$$b = 20$$

Solution   **The length of the third side is 20 units.**

You can use the Pythagorean theorem to determine if a triangle is a right triangle.

> **Converse of the Pythagorean Theorem**
>
> If a triangle has sides of length a, b, and c such that $a^2 + b^2 = c^2$, then the triangle is a right triangle with a right angle opposite side c.

Example 3

Triangle ABC has side lengths of 5 centimeters, 8 centimeters, and 10 centimeters.

Is $\triangle ABC$ a right triangle?

Strategy   **Use the converse of the Pythagorean theorem.**

Step 1   Identify the shorter sides and the longest side.

The shorter sides are 5 cm and 8 cm long. Let $a = 5$ and $b = 8$.

The longest side is 10 cm long. Let $c = 10$.

Step 2   Test the values in the equation.

$$a^2 + b^2 = c^2$$
$$5^2 + 8^2 \stackrel{?}{=} 10^2$$
$$25 + 64 \stackrel{?}{=} 100$$
$$89 \neq 100$$

So, $\triangle ABC$ is not a right triangle.

Solution   **Triangle *ABC* is <u>not</u> a right triangle.**

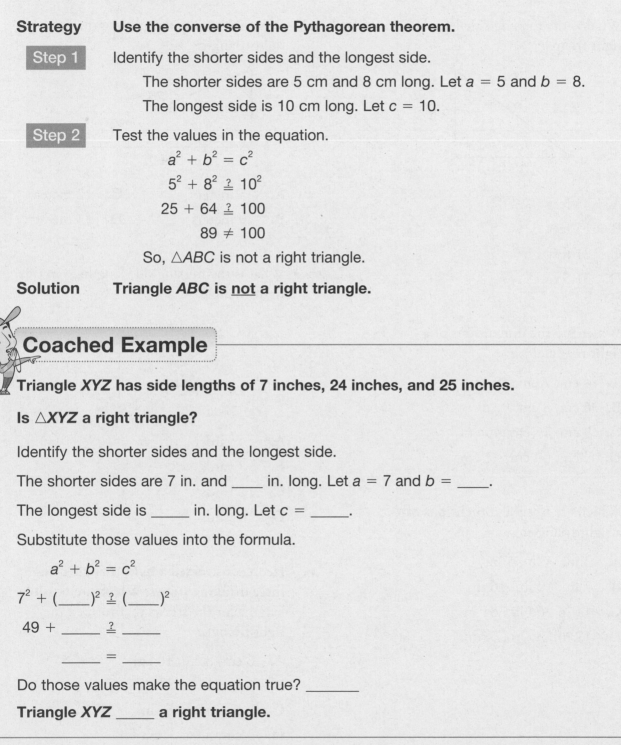

Coached Example

Triangle *XYZ* has side lengths of 7 inches, 24 inches, and 25 inches.

Is $\triangle XYZ$ a right triangle?

Identify the shorter sides and the longest side.

The shorter sides are 7 in. and _____ in. long. Let $a = 7$ and $b =$ _____.

The longest side is _____ in. long. Let $c =$ _____.

Substitute those values into the formula.

$$a^2 + b^2 = c^2$$
$$7^2 + (\underline{\quad})^2 \stackrel{?}{=} (\underline{\quad})^2$$
$$49 + \underline{\quad\quad} \stackrel{?}{=} \underline{\quad\quad}$$
$$\underline{\quad\quad} = \underline{\quad\quad}$$

Do those values make the equation true? _____

Triangle *XYZ* _____ a right triangle.

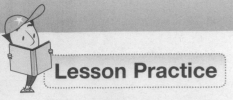

Lesson Practice

Choose the correct answer.

1. What is the missing side length, *c*, in this right triangle?

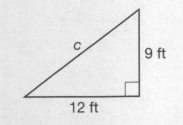

A. 3 feet

B. 15 feet

C. 21 feet

D. 31.5 feet

2. Which are the dimensions of a right triangle?

A. 4 cm, 7 cm, 10 cm

B. 6 cm, 8 cm, 9 cm

C. 8 cm, 15 cm, 40 cm

D. 9 cm, 40 cm, 41 cm

3. Which are **not** the dimensions of a right triangle?

A. 3 in., 4 in., 5 in.

B. 7 in., 10 in., 16 in.

C. 11 in., 60 in., 61 in.

D. 12 in., 35 in., 37 in.

4. What is the missing side length, *x*, in this right triangle?

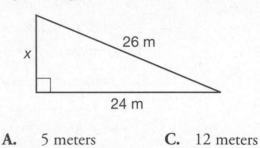

A. 5 meters   C. 12 meters

B. 10 meters   D. 43 meters

5. What is the missing side length, *x*, in this right triangle?

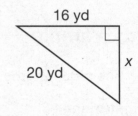

A. 4 yards

B. 8 yards

C. 12 yards

D. 16 yards

6. Jacob constructed a right triangle using three drinking straws. Which are possible lengths for the straws that formed the right triangle?

A. 6 cm, 8 cm, 10 cm

B. 5 cm, 12 cm, 15 cm

C. 3 cm, 4 cm, 6 cm

D. 2 cm, 5 cm, 7 cm

7. Ryan needs to identify a right triangle. When joined at the vertices, which set of squares below can be used to form a right triangle? (Note: Art not drawn to scale.)

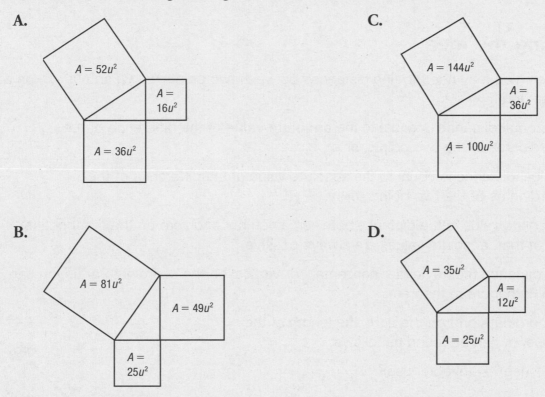

A.
$A = 52u^2$
$A = 16u^2$
$A = 36u^2$

C.
$A = 144u^2$
$A = 36u^2$
$A = 100u^2$

B.
$A = 81u^2$
$A = 49u^2$
$A = 25u^2$

D.
$A = 35u^2$
$A = 12u^2$
$A = 25u^2$

8. Marcella drew a right triangle, $\triangle BCD$, with side lengths of 9 inches, 40 inches, and 41 inches.

A. Use the converse of the Pythagorean theorem to prove that $\triangle BCD$ is a right triangle. Show your work.

B. Marcella draws another triangle with each side length double that of $\triangle BCD$. Is her new triangle also a right triangle? Show or explain your work.

Distance

Common Core State Standard:
8.G.8

Getting the Idea

Sometimes, you may need to find distances between two points (x_1, y_1) and (x_2, y_2) on a coordinate grid.

- A horizontal distance is equal to the **absolute value** of the difference of the x-coordinates of the two points, or $|x_2 - x_1|$.

- A vertical distance is equal to the absolute value of the difference of the y-coordinates of the two points, or $|y_2 - y_1|$.

Note: Absolute value is the distance between a number and zero on the number line. Because of that, absolute values are always positive.

To find a distance that is neither horizontal nor vertical on the coordinate grid, you can apply the Pythagorean theorem.

On the coordinate grid on the right, the length of the hypotenuse, d, can be found as follows:

$$(\text{leg})^2 + (\text{leg})^2 = (\text{hypotenuse})^2$$

$$(x_2 - x_1)^2 + (y_2 - y_1)^2 = d^2$$

To find d, you can take the square root of both sides of the equation.

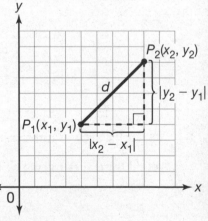

The Distance Formula

The distance d between two points (x_1, y_1) and (x_2, y_2) is:

$$d = \sqrt{(x_2 - x_1)^2 + (y_2 - y_1)^2}$$

Example 1

What is the length of $\overline{AB}$ on the coordinate grid?

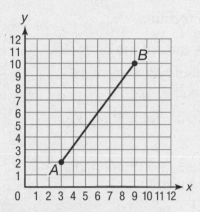

Strategy	**Use the distance formula.**

Step 1 Identify the coordinates of the endpoints.

The line segment has endpoints $A(3, 2)$ and $B(9, 10)$.

Let $(x_1, y_1) = (3, 2)$ and $(x_2, y_2) = (9, 10)$.

Step 2 Substitute the values into the distance formula and solve for d.

$$d = \sqrt{(x_2 - x_1)^2 + (y_2 - y_1)^2}$$
$$d = \sqrt{(9 - 3)^2 + (10 - 2)^2}$$
$$d = \sqrt{6^2 + 8^2}$$
$$d = \sqrt{36 + 64}$$
$$d = \sqrt{100}$$
$$d = 10$$

Solution The length of $\overline{AB}$ is 10 units.

The distance between two points is not always a whole number of units. Sometimes, the distance formula yields an irrational number. If this happens, you can estimate the number of units to the nearest tenth.

Example 2

What is the length of $\overline{CD}$?

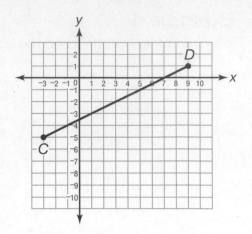

Strategy **Use the distance formula.**

Step 1 Identify the coordinates of the endpoints.

The line segment has endpoints $C(-3, -5)$ and $D(9, 1)$.

Let $(x_1, y_1) = (-3, -5)$ and $(x_2, y_2) = (9, 1)$.

Step 2 Substitute the values into the distance formula and solve for d.

$$d = \sqrt{(x_2 - x_1)^2 + (y_2 - y_1)^2}$$

$$d = \sqrt{[9 - (-3)]^2 + [1 - (-5)]^2}$$

$$d = \sqrt{12^2 + 6^2}$$

$$d = \sqrt{144 + 36}$$

$$d = \sqrt{180}$$

$$d \approx 13.4$$

Solution **The length of $\overline{CD}$ is $\sqrt{180}$ units, or approximately 13.4 units.**

Example 3

What is the length of $\overline{RS}$ with endpoints $R(-6, 2)$ and $S(-2, -3)$?

Strategy **Use the distance formula.**

Substitute the values into the distance formula and solve for d.

Let $(x_1, y_1) = (-6, 2)$.

Let $(x_2, y_2) = (-2, -3)$.

$$d = \sqrt{(x_2 - x_1)^2 + (y_2 - y_1)^2}$$

$$d = \sqrt{[-2 - (-6)]^2 + [-3 - 2]^2}$$

$$d = \sqrt{(4)^2 + (-5)^2}$$

$$d = \sqrt{16 + 25}$$

$$d = \sqrt{41}$$

$$d \approx 6.4$$

Solution **The length of $\overline{RS}$ is $\sqrt{41}$ units, or approximately 6.4 units.**

Coached Example

Which is closer to point J: point K, point L, or is each the same distance from point J?

Find the horizontal distance between points J and K by subtracting their ___-coordinates and then taking the absolute value.

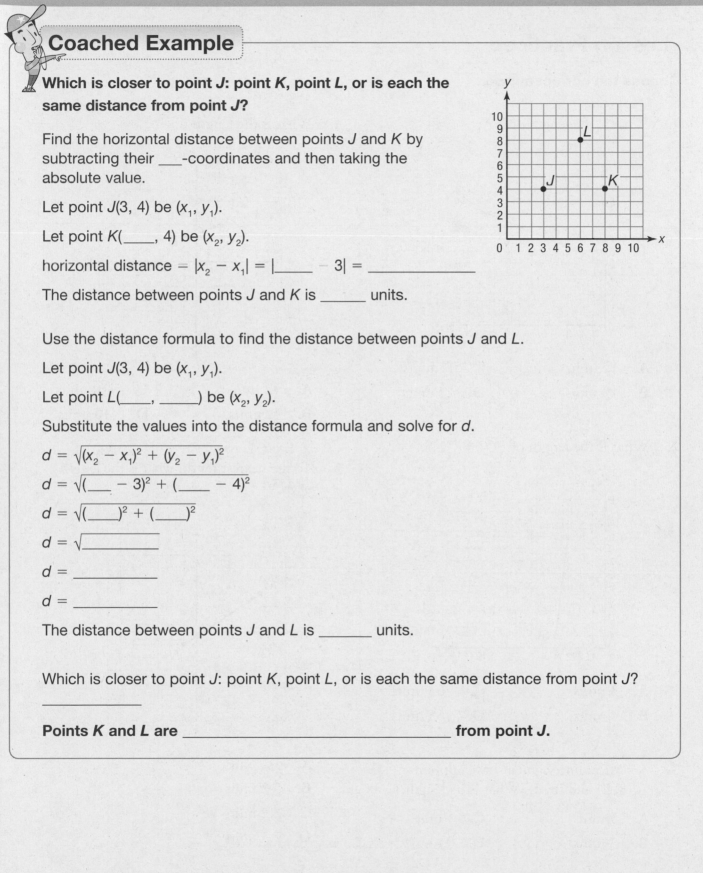

Let point $J(3, 4)$ be (x_1, y_1).

Let point $K(\underline{\quad}, 4)$ be (x_2, y_2).

horizontal distance $= |x_2 - x_1| = |\underline{\quad} - 3| = \underline{\hspace{2cm}}$

The distance between points J and K is _____ units.

Use the distance formula to find the distance between points J and L.

Let point $J(3, 4)$ be (x_1, y_1).

Let point $L(\underline{\quad}, \underline{\quad})$ be (x_2, y_2).

Substitute the values into the distance formula and solve for d.

$d = \sqrt{(x_2 - x_1)^2 + (y_2 - y_1)^2}$

$d = \sqrt{(\underline{\quad} - 3)^2 + (\underline{\quad} - 4)^2}$

$d = \sqrt{(\underline{\quad})^2 + (\underline{\quad})^2}$

$d = \sqrt{\underline{\hspace{2cm}}}$

$d = \underline{\hspace{2cm}}$

$d = \underline{\hspace{2cm}}$

The distance between points J and L is _____ units.

Which is closer to point J: point K, point L, or is each the same distance from point J?
_____

Points K and L are _____ **from point J.**

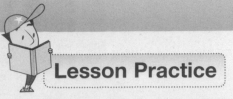

Lesson Practice

Choose the correct answer.

1. What is the length of $\overline{TV}$?

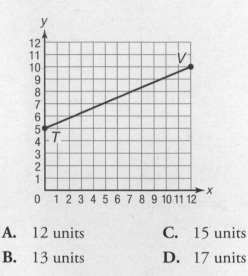

    A. 12 units   **C.** 15 units

    B. 13 units   **D.** 17 units

2. What is the length of $\overline{PQ}$?

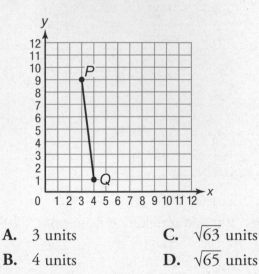

    A. 3 units   **C.** $\sqrt{63}$ units

    B. 4 units   **D.** $\sqrt{65}$ units

3. A vertical line segment has endpoints at (8, 2) and (8, 7). What is its length?

    A. 5 units   **C.** 1 unit

    B. 2 units   **D.** 0 units

4. What is the length of $\overline{AB}$?

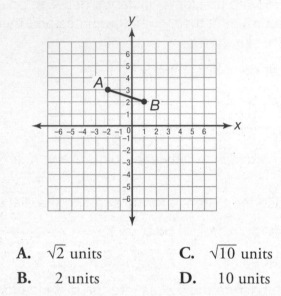

    A. $\sqrt{2}$ units   **C.** $\sqrt{10}$ units

    B. 2 units   **D.** 10 units

5. To the nearest tenth, what is the length of $\overline{MN}$?

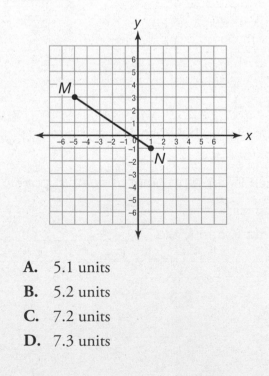

    A. 5.1 units

    B. 5.2 units

    C. 7.2 units

    D. 7.3 units

6. To the nearest tenth, what is the length of $\overline{CD}$?

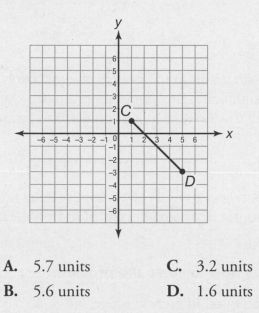

A. 5.7 units   C. 3.2 units

B. 5.6 units   D. 1.6 units

7. To the nearest tenth, what is the distance between (7, 4) and (12, 5)?

A. 4.8 units

B. 4.9 units

C. 5.0 units

D. 5.1 units

8. To the nearest tenth of a unit, what is the distance between (−6, −2) and (8, 5)?

A. 3.6 units

B. 8.5 units

C. 15.7 units

D. 15.8 units

9. Points F, G, and H are graphed below.

A. Find the distance between points G and H. Show your work.

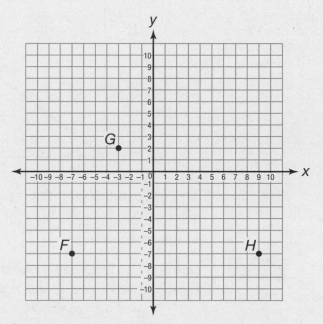

B. Is point H closer to point G than to point F, or is the distance between them the same? Explain how you know.

_____

_____

_____

_____

Apply the Pythagorean Theorem

Common Core State Standards:
8.G.7, 8.G.8

Getting the Idea

Sometimes, you may need to apply the Pythagorean theorem to solve problems involving geometric figures, real-world distances, or other situations. Drawing a diagram may help you realize that the Pythagorean theorem could be the key to solving a particular problem.

Example 1

A rectangular swimming pool has a length of 24 feet and a width of 18 feet. A hose needs to extend from the southwest corner of the pool to the northeast corner of the pool. How long does the hose need to be?

Strategy   **Draw a diagram to help you decide how to solve the problem.**

Step 1   Draw a diagram and decide what you need to find.

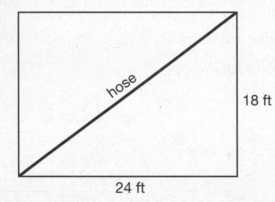

The hose in the diagram represents a diagonal of the rectangle.

The diagonal can be viewed as the hypotenuse of two right triangles in the rectangle.

Step 2   Use the Pythagorean theorem to find the diagonal of the rectangle (the hypotenuse of one of the right triangles).

$$a^2 + b^2 = c^2$$
$$24^2 + 18^2 = c^2$$
$$576 + 324 = c^2$$
$$900 = c^2$$
$$30 = c$$

Solution   **The length of the hose needs to be 30 feet.**

Example 2

A team of archaeologists fenced off an ancient ruin they are exploring. They created a grid to represent the area, so they could label the locations of several artifacts.

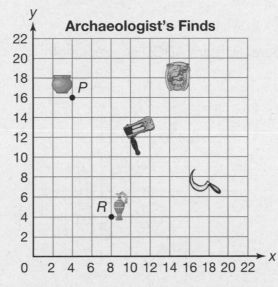

For example, they found a pot at point P and a pitcher at point R. If each unit on the grid represents 2 meters, approximately how many meters apart were the pitcher and the pot found?

Strategy   **Use the distance formula.**

Step 1   Identify the coordinates of the points.

The pot was found at $P(4, 16)$.

Let that be (x_1, y_1).

The pitcher was found at $R(8, 4)$.

Let that be (x_2, y_2).

Step 2   Substitute those values into the distance formula, and solve.

$$d = \sqrt{(x_2 - x_1)^2 + (y_2 - y_1)^2}$$
$$d = \sqrt{(8 - 4)^2 + (4 - 16)^2}$$
$$d = \sqrt{16 + 144}$$
$$d = \sqrt{160}$$
$$d \approx 12.6$$

Solution   **The pot and pitcher were found approximately 12.6 meters apart.**

Sometimes, you may need to apply more than one piece of knowledge at a time to solve a problem. For example, you may need to use what you know about rectangles and finding distances on a coordinate grid in order to solve a problem.

Example 3

Determine whether or not parallelogram *ABCD* is a rectangle.

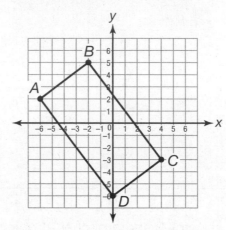

Strategy   **Use what you know about rectangles and the distance formula.**

Step 1   Think about rectangles.

The diagonals of a rectangle are congruent.

Step 2   Find the length of one of the diagonals, $\overline{AC}$.

Let $A(-6, 2)$ be (x_1, y_1) and $C(4, -3)$ be (x_2, y_2).

$d = \sqrt{(-3 - 2)^2 + [4 - (-6)]^2}$

$d = \sqrt{25 + 100}$

$d = \sqrt{125}$

Step 3   Find the length of the other diagonal, $\overline{BD}$.

Let $B(-2, 5)$ be (x_1, y_1) and $D(0, -6)$ be (x_2, y_2).

$d = \sqrt{(-6 - 5)^2 + [0 - (-2)]^2}$

$d = \sqrt{121 + 4}$

$d = \sqrt{125}$ → That's the same length as $\overline{AC}$.

Solution   **Since the diagonals are congruent, parallelogram *ABCD* must be a rectangle.**

Coached Example

Simon leans a 20-foot ladder against the side of his house so that the base of the ladder is 5 feet from the house.

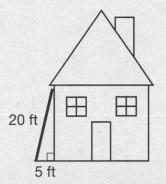

20 ft

5 ft

About how high up the side of the house does the ladder reach? Round your answer to the nearest tenth of a foot.

The ladder forms a right triangle with the house, so use the Pythagorean theorem.

The ladder is the _____ of the triangle, so let $c = $ _____.

The distance from the base of the ladder to the house is a leg, so let $a = $ _____.

Substitute those values into the formula and solve for b.

$a^2 + b^2 = c^2$

$(\underline{})^2 + b^2 = (\underline{})^2$

_____ $+ b^2 = $ _____

$b^2 = $ _____

$b = $ _____

$b \approx $ _____

The ladder reaches a height up the house of about _____ feet.

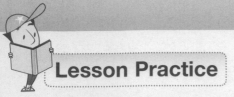

Lesson Practice

Choose the correct answer.

1. The distance across a pond cannot be directly measured. A land surveyor takes some other measurements and uses them to find *d*, the distance across the pond.

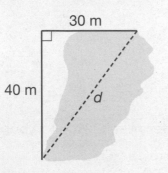

30 m

40 m

d

What is the distance across the pond?

A. 70 meters   C. 35 meters

B. 50 meters   D. 10 meters

2. On the map below, the post office is at the origin (0, 0) and each unit represents 1 km. Amy lives 6 km east and 8 km north of the post office. If she rides her bike directly from her house to the post office, how far will she ride her bike?

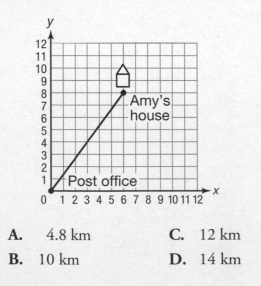

A. 4.8 km   C. 12 km

B. 10 km   D. 14 km

3. The solid lines below show the route Maddy's bus takes to school. The dashed line shows a shortcut she takes through the park when she rides her bike to school. What is the difference, in km, between the shortcut and the usual route?

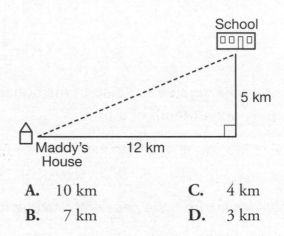

A. 10 km   C. 4 km

B. 7 km   D. 3 km

4. A quarterback throws a pass to another player.

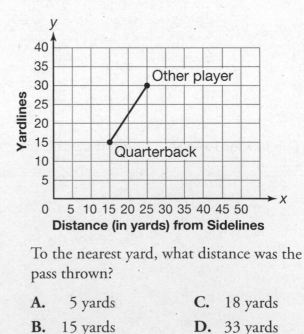

To the nearest yard, what distance was the pass thrown?

A. 5 yards   C. 18 yards

B. 15 yards   D. 33 yards

5. A guy wire is attached to the top of a 24-foot pole. The wire is attached to the ground at a point that is 10 feet from the base of the pole. What is the length of the wire?

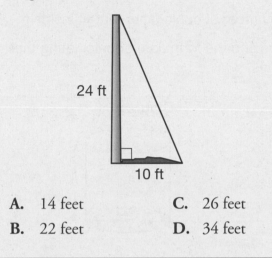

24 ft

10 ft

A. 14 feet   **C.** 26 feet

B. 22 feet   **D.** 34 feet

6. The size of a rectangular television screen is given as the length of its diagonal. The screen for a widescreen television has a length of 53 inches and a width of 30 inches. To the nearest tenth of an inch, what is the best estimate of the diagonal of that screen?

A. 83.0 inches

B. 60.9 inches

C. 43.7 inches

D. 37.1 inches

7. Rectangle *FGHJ* is shown on the grid.

A. Prove that rectangle *FGHJ* is a square. Explain or show how you proved this.

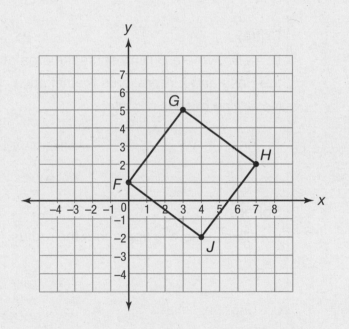

B. Find the perimeter of rectangle *FGHJ*, showing each step in the process.

Volume

Common Core State Standard:
8.G.9

Getting the Idea

The **volume** of a three-dimensional figure is the number of cubic units that fit inside it.

The table shows several three-dimensional figures and the formulas for calculating their volumes. Take the time to memorize these formulas.

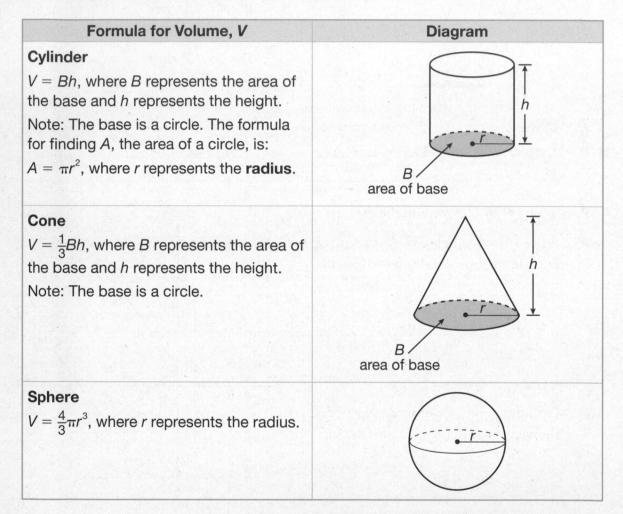

Formula for Volume, V	Diagram
Cylinder $V = Bh$, where B represents the area of the base and h represents the height. Note: The base is a circle. The formula for finding A, the area of a circle, is: $A = \pi r^2$, where r represents the **radius**.	
Cone $V = \frac{1}{3}Bh$, where B represents the area of the base and h represents the height. Note: The base is a circle.	
Sphere $V = \frac{4}{3}\pi r^3$, where r represents the radius.	

Example 1

To the nearest cubic centimeter, what is the volume of the cone on the right?

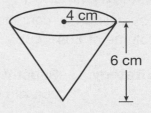

Strategy   **Use the formula for the volume of a cone.**

  Step 1   Find the approximate area of the base.

Let B represent the area of the base.

Substitute 4 for r. Use 3.14 for π.

$B = \pi r^2$

$B \approx 3.14 \times 4^2$

$B \approx 50.24 \text{ cm}^2$

  Step 2   Find the volume.

Substitute 50.24 for B and 6 for h.

$V = \frac{1}{3}Bh$

$V \approx \frac{1}{3} \times 50.24 \times 6$

$V \approx 100.48 \text{ cm}^3$

100.48 cm^3 rounds down to 100 cm^3.

Solution   **The volume is approximately 100 cubic centimeters.**

The area of the base of a cone, B, is equal to πr^2.

So, you can write the formula for finding the volume of a cone as:

$V = \frac{1}{3}Bh$ or $V = \frac{1}{3}\pi r^2 h$.

Notice that in the second formula, πr^2 is substituted for B.

For the same reason, you can use either of the formulas below to find the volume of a cylinder:

$V = Bh$ or $V = \pi r^2 h$.

Example 2

The cylindrical can shown on the right has a height of 24 inches and a volume of 864π cubic inches. What is its radius, r?

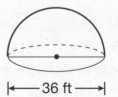

24 in.

Strategy   **Substitute the known values into the formula for the volume of a cylinder. Solve for r.**

Step 1   Substitute the known values into the formula.

Substitute 864π for the volume, V. Substitute 24 for h.

$$V = \pi r^2 h$$
$$864\pi = \pi \times r^2 \times 24$$
$$864\pi = 24\pi r^2$$

Step 2   Solve for r.

$864\pi = 24\pi r^2$	Divide both sides by π.
$\dfrac{864\pi}{\pi} = \dfrac{24\pi r^2}{\pi}$	
$864 = 24r^2$	Divide both sides by 24.
$36 = r^2$	Take the square root of both sides.
$6 = r$	

Solution   **The radius of the cylinder is 6 inches.**

Sometimes, in a diagram, the **diameter** of a figure is given instead of the radius. The radius of a figure is equal to half its diameter.

Example 3

A museum wants to build a small planetarium. The dome for the planetarium will be a hemisphere with a diameter of 36 feet.

What will be the approximate volume of the dome to the nearest cubic foot?

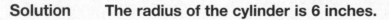

|← 36 ft →|

Strategy   **Use the formula for the volume of a sphere to find the volume of the hemisphere.**

Step 1   Determine the radius.

The radius, r, is half the diameter, d, and $d = 36$ feet.

$$r = \frac{d}{2} = \frac{36}{2} = 18 \text{ ft}$$

Step 2   Decide what a hemisphere is.

A hemisphere is a half-sphere.

So, multiply the formula for finding the volume of a sphere by $\frac{1}{2}$.

Step 3 Substitute the known values into the formula.

Substitute 18 for r. Use 3.14 for π.

$$V = \frac{1}{2} \times \frac{4}{3}\pi r^3$$

$$V \approx \frac{1}{2} \times \frac{4}{3} \times 3.14 \times 18^3$$

Step 4 Calculate the approximate volume.

$$V \approx \frac{1}{2} \times \frac{4}{3} \times 3.14 \times 18^3$$

$$V \approx \frac{1}{2} \times \frac{4}{3} \times 3.14 \times 5,832$$

$$V \approx 12,208.32$$

12,208.32 ft^3 rounds down to 12,208 ft^3.

Solution The dome will have a volume of approximately 12,208 cubic feet.

Coached Example

To the nearest tenth of a cubic centimeter, what is the volume of the soup can below?

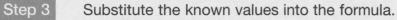

The diameter, d, is _____ cm. Calculate the radius, r.

$r = \frac{d}{2} =$ _____

Substitute _____ for the height, h, and _____ for the radius, r, into the formula.

Use 3.14 for π and calculate the volume.

$V = \pi r^2 h$

$V \approx 3.14 \times ($_____$)^2 \times$ _____

$V \approx 3.14 \times$ _____ $\times$ _____   Evaluate the exponent.

$V \approx$ _____   Multiply.

That number rounded to the nearest tenth is _____.

The volume of the can is approximately _____ cubic centimeters.

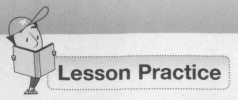

Lesson Practice

Choose the correct answer.

1. What is the approximate volume of this cylinder?

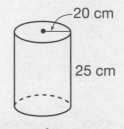

-20 cm

25 cm

A. 3,140 cm³

B. 7,850 cm³

C. 10,000 cm³

D. 31,400 cm³

2. What is the approximate volume of this cone?

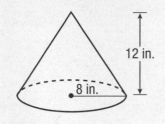

12 in.

8 in.

A. 100 in.³

B. 256 in.³

C. 804 in.³

D. 2,412 in.³

3. A ball shaped like a sphere has a radius of 15 centimeters. Which is closest to the amount of air needed to fill the ball?

A. 42,390 cm³

B. 14,130 cm³

C. 2,826 cm³

D. 942 cm³

4. The cylinder below is a storage tank for heating fuel. Which is closest to the amount of fuel the tank can hold?

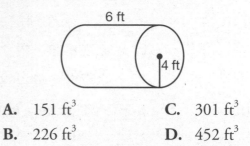

6 ft

4 ft

A. 151 ft³   **C.** 301 ft³

B. 226 ft³   **D.** 452 ft³

5. Which is closest to the volume of a bowling ball with a diameter of 21.8 centimeters?

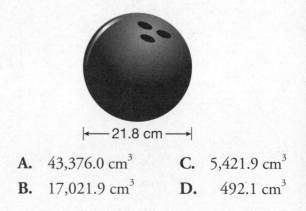

←—21.8 cm—→

A. 43,376.0 cm³   **C.** 5,421.9 cm³

B. 17,021.9 cm³   **D.** 492.1 cm³

6. The cylinder shown below has a volume of 300π cubic meters. What is its height?

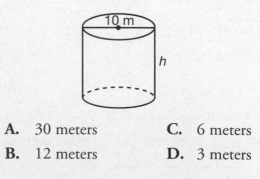

10 m

h

A. 30 meters   **C.** 6 meters

B. 12 meters   **D.** 3 meters

7. Daniel is going to melt some wax and pour it into a mold shaped like the cone shown below.

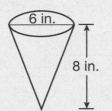

Which is closest to the volume of the cone?

A. 75 in.^{3} C. 226 in.^{3}

B. 151 in.^{3} D. 301 in.^{3}

8. What is the volume of this hemisphere?

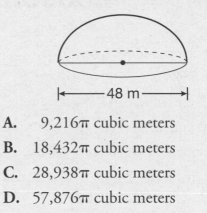

A. 9,216π cubic meters

B. 18,432π cubic meters

C. 28,938π cubic meters

D. 57,876π cubic meters

9. A company manufactures cylindrical tanks that store water. Tank A shown below is the company's best-selling tank.

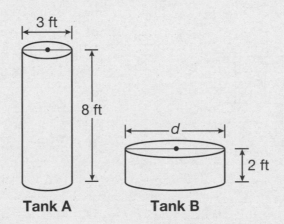

Tank A   **Tank B**

A. Calculate the exact volume of Tank A, showing each step in the process. Do **not** use an approximation for π.

B. The company wants to manufacture a second tank that is shorter than Tank A but has the same volume. If Tank B has a height of 2 feet and a volume equal to Tank A, what will be its diameter? Show or explain your work below.

Domain 4: Cumulative Assessment for Lessons 24–32

1. Which are the dimensions of a right triangle?

   A. 12 cm, 35 cm, 37 cm

   B. 10 cm, 15 cm, 20 cm

   C. 8 cm, 12 cm, 16 cm

   D. 3 cm, 4 cm, 7 cm

2. If square D has an area of 225 square centimeters and square E has an area of 81 square centimeters, what is the area of square F?

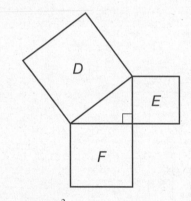

   A. 144 cm^2

   B. 153 cm^2

   C. 169 cm^2

   D. 306 cm^2

Use the diagram for questions 3 and 4.

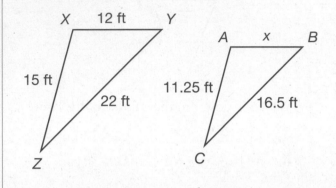

3. Triangle XYZ is similar to *triangle ABC* because two pairs of corresponding sides have proportional lengths and the pair of angles between them are congruent. Which of the following is **not** true and would **not** help to prove this?

   A. $\frac{XZ}{AC} = \frac{YZ}{BC}$

   B. $\frac{XZ}{AC} = \frac{4}{3}$ and $\frac{YZ}{BC} = \frac{4}{3}$

   C. $m\angle X = m\angle C = 20°$

   D. $m\angle Z = m\angle C = 20°$

4. What is the value of x?

   A. 9 feet

   B. 12 feet

   C. 16 feet

   D. 20 feet

5. The 14-foot ladder below is leaning against a wall. Its base is 3 feet from the wall.

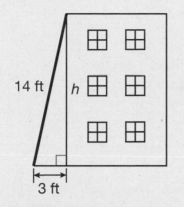

14 ft
h
3 ft

To the nearest tenth of a foot, how high above the ground does the ladder reach?

A. 11.0 feet

B. 13.7 feet

C. 14.3 feet

D. 18.7 feet

6. The cylindrical fish tank shown below has a height of 25 inches and a volume of $5,625\pi$ cubic inches. What is the diameter of the fish tank?

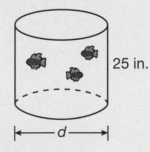

25 in.

d

A. 15 inches

B. 30 inches

C. 60 inches

D. 225 inches

7. The measure of an exterior angle of a triangle is $3x - 5$ degrees. The measure of the corresponding interior angle is $5x + 17$ degrees. What is the measure of the exterior angle?

A. 21°

B. 32°

C. 58°

D. 122°

8. A triangle is graphed on the coordinate grid below.

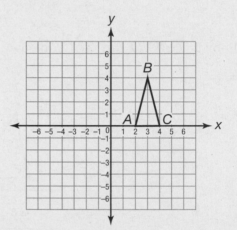

What are the coordinates of the image of this triangle after a 180° rotation about the origin?

A. $A'(-2, 0), B'(-3, -4), C'(-4, 0)$

B. $A'(0, -2), B'(-4, -3), C'(0, -4)$

C. $A'(0, 2), B'(4, 3), C'(0, 4)$

D. $A'(2, 0), B'(3, -4), C'(4, 0)$

9. What is the length of $\overline{AB}$, in units?

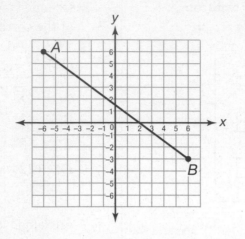

_____

10. Look at $\triangle MNO$ and $\triangle PQR$ below.

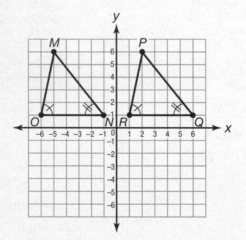

A. Explain how you could prove that $\triangle MNO$ is congruent to $\triangle PQR$.

_____

_____

_____

B. List all the pairs of corresponding, congruent sides and corresponding, congruent angles for these triangles.

_____

_____

_____

Domain 5

Statistics and Probability

Domain 5: Diagnostic Assessment for Lessons 33–36

Lesson 33 Scatter Plots
8.SP.1

Lesson 34 Trend Lines
8.SP.1, 8.SP.2

Lesson 35 Interpret Linear Models
8.SP.3

Lesson 36 Patterns in Data
8.SP.4

Domain 5: Cumulative Assessment for Lessons 33–36

Domain 5: Diagnostic Assessment for Lessons 33–36

Use the scatter plot for questions 1 and 2.

The scatter plot below compares the number of hours of sleep that students got the night before an oral report to their grades on the report.

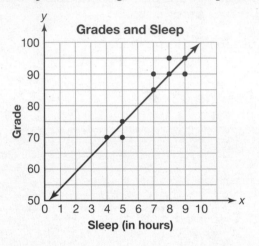

1. What type of association, if any, is shown by the scatter plot?

   A. negative association

   B. no association

   C. nonlinear association

   D. positive association

2. Which statement is true about the line of best fit drawn above?

   A. The line comes close to most points, so it is a very good model for the data.

   B. The line shows the correct association, but does not come close to most points.

   C. The scatter plot shows no association, so a line should not be used to model the data.

   D. The data do not resemble a straight line, so a nonlinear model would be better for these data.

Use the scatter plot for questions 3 and 4.

The scatter plot below compares the price of each pair of jeans sold at a boutique to the number of pairs sold last month.

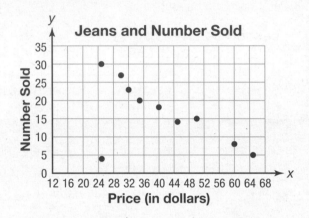

3. If the data above contain an outlier, which coordinates best represent it?

   A. (25, 4)

   B. (45, 14)

   C. (65, 5)

   D. There is no outlier for these data.

4. Which best describes the association shown by the scatter plot?

   A. positive, linear association

   B. negative, linear association

   C. nonlinear association

   D. no association

5. Which scatter plot shows a nonlinear association for the data?

A.

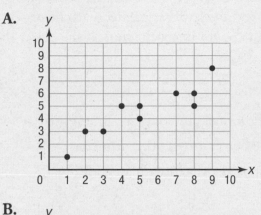

B.

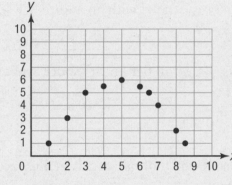

C.

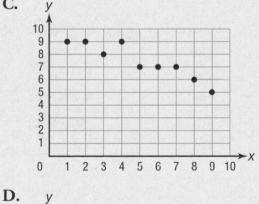

D.

Use the table for questions 6 and 7.

A survey of voters explored the relationship between the ages of voters and whether or not they support building skateboard ramps at a local park.

Age	In Favor	Opposed	Total
18–37	40	10	50
38–57	27	23	50
58 and over	6	44	50
Total	73	77	150

6. What percentage of surveyed voters in the 58 and over age group are opposed to building skateboard ramps?

A. 88%

B. 75%

C. 44%

D. 29%

7. Which is **not** a reasonable interpretation of the data?

A. Ignoring age groupings, about the same number of total voters favor building skateboard ramps as oppose it.

B. Voters in the 18–37 age group are much more likely to favor building skateboard ramps than to oppose it.

C. Voters in the 38–57 age group are much more likely to favor building skateboard ramps than to oppose it.

D. Voters in the 58 and over age group are much more likely to oppose building skateboard ramps than to favor it.

Use the information for questions 8 and 9.

Ariel bought potted flowers to grow indoors. She collected data last year and discovered that placing the flowers under a heat lamp once a day for 30 minutes was associated with increased growth. The equation $y = 1.25x$ shows y, the increase in growth in centimeters, if she places the flowers under the heat lamp for x weeks.

8. What association is shown by the linear model?

   A. For every week that passes, the plant shows additional growth of 1.25 centimeters.

   B. For every 1.25 hours under the heat lamp, the flower grows 30 centimeters.

   C. Each additional half hour under the heat lamp is associated with an additional 1.25 centimeters of flower growth.

   D. Each additional 1.25 hours under the heat lamp is associated with an additional 0.5 centimeter of flower growth.

9. Ariel places a flower under the heat lamp for 8 weeks. Based on her linear model, what should she expect the flower's total additional growth to be?

   _____

10. Ten students in a class—5 boys and 5 girls—were asked which of two movie genres (action or comedy) is their favorite. The results are shown below.

Student	Favorite Movie Genre
boy	action
girl	comedy
girl	comedy
boy	comedy
boy	action

Student	Favorite Movie Genre
girl	action
girl	comedy
boy	comedy
girl	comedy
boy	action

   A. Complete the two-way table below to show these results.

	Action	Comedy	Total
Boy			
Girl			
Total			

   B. Create a second two-way table that shows relative frequencies for the table from Part A.

	Action	Comedy	Total
Boy			
Girl			
Total			

Scatter Plots

Common Core State Standard:
8.SP.1

Getting the Idea

A **scatter plot** is a graph in which ordered pairs of data are plotted. You can use a scatter plot to determine if a relationship, or an association, exists between two sets of data. There are different kinds of associations.

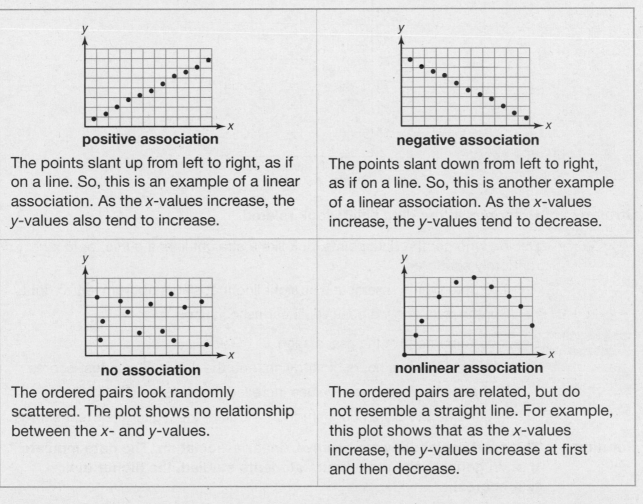

positive association

The points slant up from left to right, as if on a line. So, this is an example of a linear association. As the x-values increase, the y-values also tend to increase.

negative association

The points slant down from left to right, as if on a line. So, this is another example of a linear association. As the x-values increase, the y-values tend to decrease.

no association

The ordered pairs look randomly scattered. The plot shows no relationship between the x- and y-values.

nonlinear association

The ordered pairs are related, but do not resemble a straight line. For example, this plot shows that as the x-values increase, the y-values increase at first and then decrease.

An association does not have to be true for every pair of values in a scatter plot. It should be true for most of the data points. Look at how the data cluster together to help you decide.

Example 1

The scatter plot shows the number of hours that students studied for a test and their scores on the test. Does the scatter plot show an association between the number of hours studied and test scores? If so, describe the association.

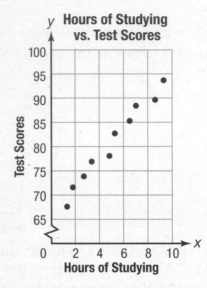

Strategy   **Determine if the data points look related.**

Step 1   Decide whether the data points look like a straight line, a curve, or look randomly scattered.

      The data points resemble a straight line that slants up from left to right.

      So, the data show a positive, linear association.

Step 2   Describe and interpret the association.

      The x-axis shows hours of studying, and the y-axis shows test scores.

      The plot shows that as x-values increase, y-values tend to increase.

      In general, the more hours students studied, the higher their test scores.

Solution   **The scatter plot shows a positive, linear association. The data indicate that, in general, the more hours students studied, the higher their test scores.**

Notice that the scatter plot in Example 1 has a squiggle mark on the y-axis. This indicates a break in the axis, where the scale jumps from 0 to 65. It can be useful to include a break in an axis for certain data.

Example 2

The data below show the average quarterly stock price for a company during its first three years in business.

Average Stock Prices

Quarter	1	2	3	4	5	6	7	8	9	10	11	12
Price per Share	$2	$4	$8	$11	$13	$14	$13	$12	$9	$8	$5	$1

Create a scatter plot for these data and identify the association, if any.

Strategy **Choose a scale. Then title and label a grid. Plot each point on the grid. Identify the association, if any.**

Step 1 Choose a title and axis labels for the grid.

Give the graph the same title as the table.

There are four quarters in one year. So, a quarter is a unit of time.

The other variable is price per share.

The quarter cannot be affected by the price per share, but the price can change each quarter.

So, the quarter is the independent variable. Label the x-axis "Quarter."

The price per share is the dependent variable. Label the y-axis "Price Per Share."

Step 2 Choose a range and scale for each axis.

Since the greatest x-value in the table is 12, use 0 to 12 with intervals of 1.

The greatest y-value in the table is 14, so a scale of 0 to 16, with intervals of 2, is a good scale.

Step 3 Plot each data point on the grid.

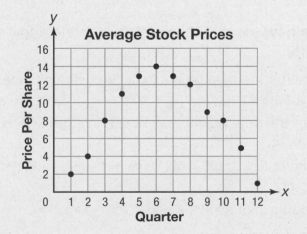

Step 4 Analyze the data for an association.

The data points resemble a U-shaped curve.

Since the data points resemble a curve instead of a straight line, this shows a nonlinear association.

Solution **The scatter plot is shown in step 3. It shows a nonlinear association.**

When you look at a set of data on a scatter plot, you may notice that some data points do not give a good indication of the association shown by almost all the other data points. An **outlier** is a data point with values that are significantly different from the other data points in the set. It is often helpful to ignore an outlier when determining the association shown by a scatter plot.

Example 3

The manager of an outdoor theater space wanted to find out if ticket prices affect the number of people who attend a concert. The scatter plot below shows the data she collected.

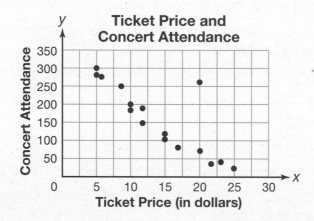

What association is shown by the scatter plot? Did you exclude any outliers when determining that association?

Strategy **Determine how most of the points are related. Identify and exclude any outliers.**

Step 1 Decide whether the data points look like a straight line, a curve, or look randomly scattered.

Most of the data points resemble a straight line that slants down from left to right.

Only the outlier at (20, 260) does not seem to fit near such a line. Exclude that point.

The scatter plot shows a negative association.

Step 2 Describe and interpret the association.

The *x*-axis shows ticket prices and the *y*-axis shows concert attendance.

The plot shows that, in general, as ticket prices increase, concert attendance decreases.

The outlier at (20, 260) is an exception. It shows that one concert that had $20 tickets still had high attendance. Perhaps that was just a very popular concert.

Solution The scatter plot shows a negative association. The outlier at (20, 260) was excluded because it did not seem to fit with the other data.

Coached Example

The scatter plot shows the heights, in inches, of a class of students and their scores on a test. What type of association, if any, is shown by the scatter plot?

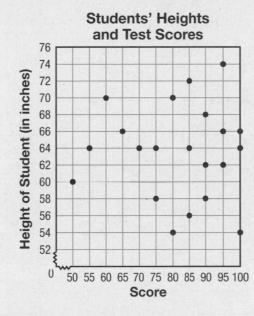

Do most of the data points resemble a straight line? _____

Do most of the data points resemble a curve? _____

Do the data points appear to be randomly scattered? _____

Since the data points appear _____, this scatter plot shows that there is _____ association between students' heights and their test scores.

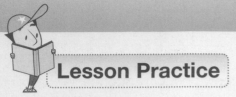

Lesson Practice

Choose the correct answer.

1. Which scatter plot shows a positive association for the data?

2. Which scatter plot shows a nonlinear association for the data?

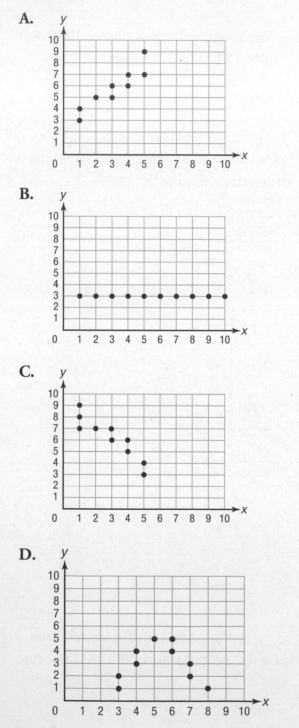

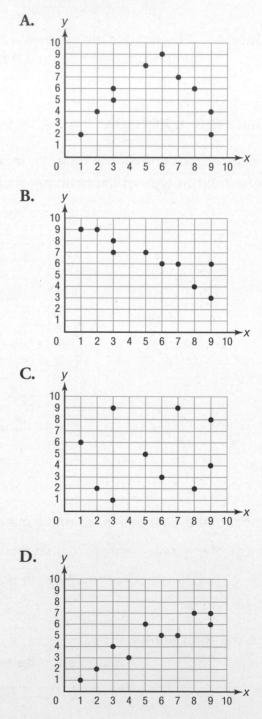

Use the scatter plot for questions 3 and 4.

The scatter plot below compares the weights of laptop computers to their prices.

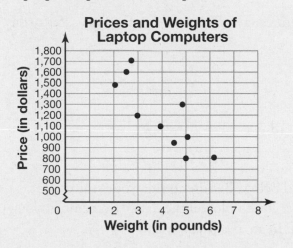

3. If the data above contain an outlier, which coordinates best represent it?

   A. (2.5, 1600)

   B. (5, 800)

   C. (6.25, 800)

   D. There is no outlier for these data.

4. Which best describes the association shown by the scatter plot?

   A. positive, linear association

   B. negative, linear association

   C. nonlinear association

   D. no association

5. Which best describes the association between the runs scored per game by the Tigers and the runs scored per game by the Lions?

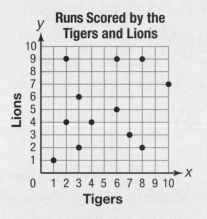

   A. negative association

   B. no association

   C. nonlinear association

   D. positive association

6. The scatter plot compares the amount of fertilizer used per acre and the corn yield for that acre. Which best describes the association?

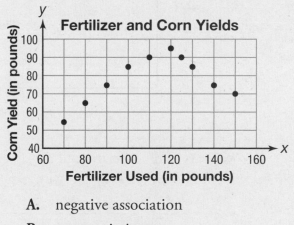

   A. negative association

   B. no association

   C. nonlinear association

   D. positive association

7. Paula has a lemonade stand, which she operates rain or shine. The table below shows the daily high temperature and the number of cups of lemonade she sold each day last week. It rained on Saturday, but it was sunny every other day.

Day	Sunday	Monday	Tuesday	Wednesday	Thursday	Friday	Saturday
Daily High Temperature (in °F)	84	90	92	87	87	95	93
Number of Cups Sold	10	30	36	18	20	48	5

A. Create a scatter plot of these data on the grid below. Be sure to title your scatter plot, label each axis, and choose a scale that allows you to plot all the data. (Remember, you can draw a squiggle mark to indicate a break in an axis if you wish.)

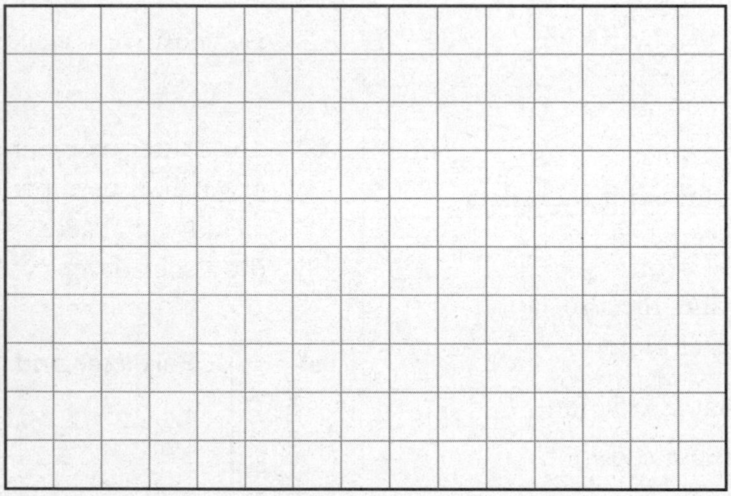

B. What type of association, if any, is shown by the scatter plot? If you excluded any outliers, identify them and explain why.

_____

_____

_____

Trend Lines

Common Core State Standards:
8.SP.1, 8.SP.2

Getting the Idea

If there is a linear association between the data on a scatter plot, you can draw a **line of best fit** to show the general trend of the data. This line is also called a **trend line**. There is usually no line that will fit every data point exactly, but the line should be as close to as many of the points on the scatter plot as possible, with about as many points above the line as below it and including at least a few points on the line.

Example 1

The scatter plot shows the heights and weights of players on a basketball team. Draw a line of best fit for these data and discuss how well the line you drew models the trend of the data.

Strategy   **Draw a line of best fit. Then describe the general trend.**

Step 1   Draw a line of best fit to show the general trend of the data.

Try to draw a line that has about as many points above it as below it.

Step 2   Analyze your line of best fit.

The line shows a positive association. So, the taller a player is, the heavier his weight is.

The line includes two of the data points and has three points above it and four points below it. The points that do not lie on the line are not very close to the line.

So, it is a decent model, but not a great model, for these data.

Solution   **The line of best fit drawn in Step 2 shows the general trend that, the taller the player, the greater the weight.**

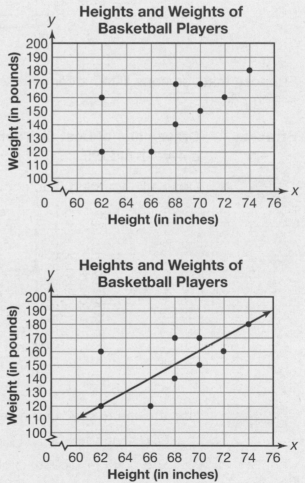

Heights and Weights of Basketball Players

Drawing a line of best fit also helps you make predictions based on the scatter plot.

Example 2

The scatter plot below shows the ages of 16 cars listed for sale online and their selling prices.

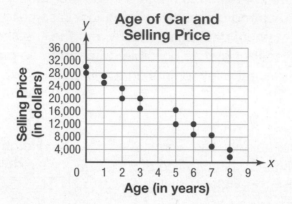

Suppose that someone listed a car for sale that is 4 years old. Make a prediction about the selling price.

Strategy **Draw a line of best fit. Then use the line to estimate the selling price for a 4-year-old car.**

Step 1 Draw a line of best fit to show the general trend of the data.

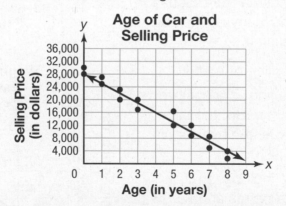

Step 2 Use the line to predict the selling price for a 4-year-old car.

Find 4 on the *x*-axis. Follow it up to the line.

The point (4, 16000) is on the line. So, a good prediction is $16,000.

Solution **According to the line of best fit, a 4-year-old car would have a selling price of about $16,000.**

Coached Example

The scatter plot below shows the relationship between the high temperature in degrees Fahrenheit (°F) and the elevation for 16 towns within a degree of latitude on the same day. Predict the high temperature for a town that has an elevation of 350 feet.

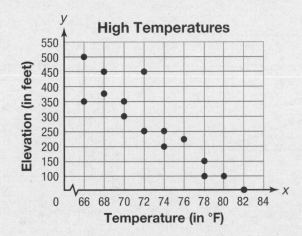

On the scatter plot above, draw a line of best fit to show the general trend of the data.

Try to draw a line that has about as many points above it as below it.

Find 350 on the y-axis. Move your finger to the right until you reach the line of best fit.

What temperature corresponds to 350 feet? _____

I predict the temperature will be about _____°F at an elevation of 350 feet.

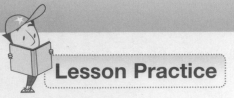

Lesson Practice

Choose the correct answer.

1. Which of these lines best fits the given data?

   A.

   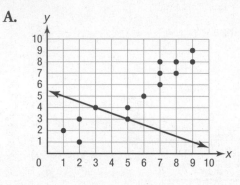

   B.

   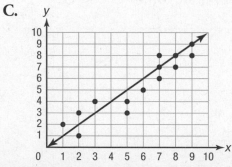

   C.

   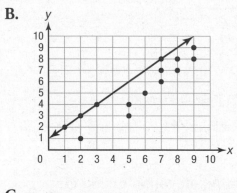

   D.

   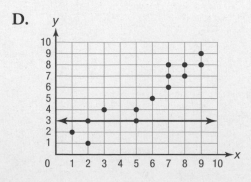

2. The scatter plot compares the number of grams of fat to the number of calories in some foods. A line of best fit has been drawn for these data.

   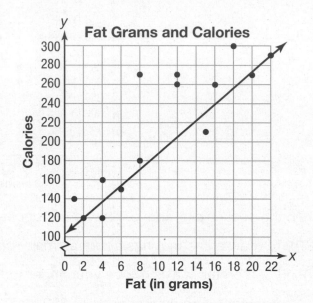

   Which statement is true about the line of best fit drawn above?

   A. The line comes close to most points, so it is a very good model for the data.

   B. The line shows the correct association, but it does not come close to most points.

   C. The scatter plot shows no association, so a line should not be used to model the data.

   D. The data do not resemble a straight line, so a nonlinear model would be better for these data.

3. The scatter plot shows the airfares paid and the distances that customers traveled. A line of best fit has been drawn for these data.

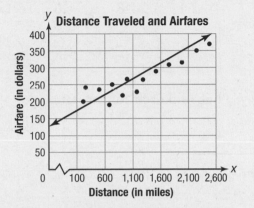

Based on the data in the scatter plot, which is the best prediction for the cost of a 100-mile trip?

A.	$75	**C.**	$175
B.	$100	**D.**	$250

4. The scatter plot below compares students' heights to the number of text messages they send daily. What would be the best prediction of the number of text messages sent by a student who is 68 inches tall?

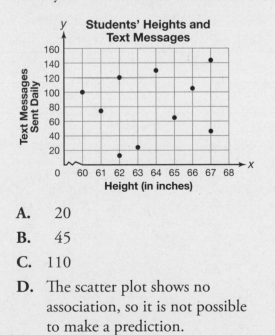

A. 20

B. 45

C. 110

D. The scatter plot shows no association, so it is not possible to make a prediction.

5. The scatter plot below shows the attendance at a sports team's away games. What would be the best prediction of the number of fans who would attend an away game if they had to travel 33 miles?

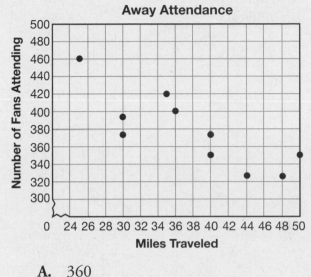

A. 360

B. 400

C. 440

D. The scatter plot shows no association, so it is not possible to make a prediction.

6. The table below shows the heights and arm spans of 9 people.

Heights and Arm Spans

Height (in meters)	1.5	1.7	2.0	1.6	1.7	2.1	1.8	1.6	1.8
Arm Span (in meters)	1.4	1.6	1.9	1.6	1.7	2.1	1.7	1.5	1.8

A. Make a scatter plot of the data above. Then draw a line of best fit for the data.

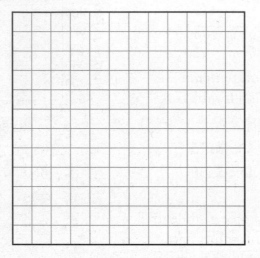

B. Predict the arm span of a person whose height is 1.3 meters. Explain how you made your prediction.

_____

_____

_____

Common Core State Standard:
8.SP.3

Interpret Linear Models

Getting the Idea

As you have already learned, data can sometimes be modeled by a straight line called a line of best fit. This line is a linear model. Think of it as using a linear function to represent the trend of the data points.

A linear function can be represented in different ways—in a table, by a graph, or as an equation. You can also use different representations to show the trend in a set of data. You could draw a scatter plot and a line of best fit, or you could use an equation or other representation.

Example 1

Elena bought small potted plants and transferred them to her garden. She collected data last year and discovered that turning on the sprinkler for a short time each night was associated with increased plant growth. The equation $y = 0.5x$ shows y, the increase in plant growth in centimeters, if she turns on the sprinkler for x nights.

What is the increase in plant growth each night that she turns on the sprinkler? What is the increase if she turns on the sprinkler each night for a week? for 2 weeks?

Strategy   **Interpret the equation. Then use it to calculate the increases in plant growth over time.**

> **Step 1**   What is the increase in plant growth each night?
>
> Since x stands for the number of nights, substitute 1 for x in the equation.
>
> $y = 0.5x = 0.5(1) = 0.5$
>
> Turning on the sprinkler is associated with a 0.5-centimeter increase in plant growth each night.

> **Step 2**   What is the increase in plant growth if Elena turns on the sprinkler for 1 or 2 weeks?
>
> One week = 7 days, so $y = 0.5(7) = 3.5$ cm.
>
> Two weeks = 14 days, so $y = 0.5(14) = 7.0$ cm.

Solution   **Turning on the sprinkler is associated with an additional 0.5 centimeter of daily plant growth. If Elena turns on the sprinkler each night for 1 week, her plants will grow an additional 3.5 centimeters. They will grow 7 additional centimeters if she does so for 2 weeks.**

The equation $y = 0.5x$ in Example 1 is a direct proportion in the form $y = mx$. The value of m, 0.5, represents the increase in plant growth per day.

When a linear model is used to represent what has been learned from collected pairs of data, you can look at the slope and intercepts to help you understand the data. You know already that the y-intercept is the point at which the graph crosses the y-axis. The **x-intercept** may also be important. It shows the point at which the graph crosses the x-axis.

Example 2

Last year, when Dayshawn bought a new computer, he collected pairs of data in a scatter plot to help him understand how the value of a computer changes over time. He used the scatter plot on the left to create the linear model on the right.

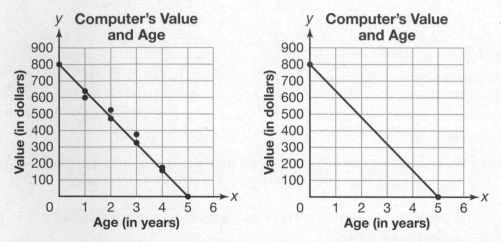

Identify the slope and intercepts of Dayshawn's graph. What do they represent in this problem? Is his linear model a perfect predictor of his computer's value at any given time?

Strategy   **Identify the slope and intercepts and what they represent in the problem.**

    Step 1   Identify the slope and what it represents.

Use the points (0, 800) and (5, 0).

$$m = \frac{0 - 800}{5 - 0}$$

$$m = -\frac{800}{5}$$

$$m = -160$$

The slope compares value in dollars to age in years. So, it shows that a computer's value decreases by approximately $160 per year.

| Step 2 | Identify the intercepts and what they represent. |

The *y*-intercept at (0, 800) shows an initial computer value of $800.

The *x*-intercept at (0, 5) shows that after 5 years a computer is not worth much, if anything. This type of information could help Dayshawn decide whether to upgrade a current computer or buy a new one.

| Step 3 | Evaluate the accuracy of the linear model. |

It is important to remember that a linear model, while helpful, is never perfect.

For example, if Dayshawn tries to sell his computer when it is 2 years old, there is no guarantee that he will be able to find someone willing to pay $480 for it.

Solution   **The slope shows that the age of a computer is associated with a decrease of about $160 per year in value. The intercepts show an initial computer value of $800 and that after 5 years, a computer will not have much, if any, value.**

Coached Example

Raquel determines that the number of points she scores in a basketball game is related to the number of minutes she plays in a game. The table below represents several ordered pairs in her linear model.

Raquel's Basketball Games

Minutes Played (x)	0	5	10	15	20	25
Points Scored (y)	0	3	6	9	12	15

What is the slope, and what does it represent in the problem?

Find the slope. Use the points (_____, 3) and (10, 6).

$$m = \frac{6 - 3}{10 - \underline{}}$$

$$m = \underline{}$$

The slope compares points scored to minutes played.

So, her model shows that for each additional _____ minutes she plays, she tends to score _____ additional points. However, this model cannot predict exactly how many points Raquel will score if she plays for x minutes in any one game.

Lesson Practice

Choose the correct answer.

Use the information for questions 1 and 2.

Last year, Mr. Wu collected data to determine the association between the number of hours that students spent on a research project and their final grades on the project. The equation $y = 60 + 5x$ was a good linear model for determining y, a student's final grade on the project, after working on it for x hours.

1. Mr. Wu assigned the same project this year. If a student spends 6 hours working on the project, which grade does the linear model predict the student will receive?

   A. 30   **C.** 90

   B. 70   **D.** 93

2. What association is shown by the linear model?

   A. Every 5 hours of work on the project increases a student's grade by 5 points.

   B. Every 5 hours of work on the project increases a student's grade by 60 points.

   C. Each additional hour of work on the project increases a student's grade by $\frac{1}{2}$ point.

   D. Each additional hour of work on the project increases a student's grade by 5 points.

Use the information for questions 3 and 4.

Macy determined that the number of additional blooms produced by a flower bush is related to the number of times she adds special plant food to its soil over the course of the season. The table below represents several ordered pairs in her linear model.

Times Added	Number of Additional Blooms
0	0
1	2
2	4
3	6
4	8

3. What does the slope of the linear model show?

   A. It shows that each time the plant food is added, the bush produces exactly 2 blooms.

   B. It shows that each time the plant food is added, the bush produces 2 more blooms than it ordinarily does.

   C. It shows that each time the plant food is added, the bush produces 2 fewer blooms than it ordinarily does.

   D. It shows that adding plant food is not associated with an increase in the number of blooms.

4. If Macy adds special plant food to the soil 6 times this season, how many additional blooms would you predict the bush to produce?

A. 6

B. 8

C. 12

D. 14

Use the graph for questions 5 and 6.

A video store owner determined that the number of times a particular movie is rented is associated with the number of weeks since it was first released. The linear model below is a good representation of this association.

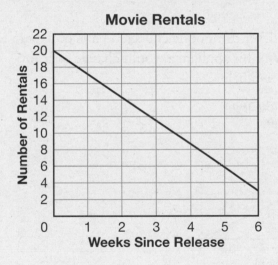

5. What does the *y*-intercept represent?

A. It shows that during the week a movie is released, it is rented about 20 times.

B. It shows that 6 weeks after a movie is released, it is rented about 3 times.

C. It shows that 3 weeks after a movie is released, it is rented about 6 times.

D. It shows that 20 weeks after a movie is released, it is rented 0 times.

6. What does the slope represent?

A. It shows that the number of times a movie is rented increases by 3 each week after its release.

B. It shows that the number of times a movie is rented decreases by 3 each week after its release.

C. It shows that a movie is rented 3 times as often each week after its release.

D. It shows that there is no association between the number of times a movie is rented and how many weeks have passed since its release.

Use the information for questions 7 and 8.

Last semester, Ms. Allen collected data to determine the association between her students' television viewing the day before an exam and their final grades on the exam. The equation $y = 100 - 8x$ was a good linear model for determining y, a student's final grade on the exam, after watching TV for x hours the night before.

7. If a student spends 4.5 hours watching television the night before the next exam, which grade does the linear model predict the student will receive?

   A. 95   **C.** 64

   B. 88   **D.** 36

8. What association is shown by the linear model?

   A. Watching one additional hour of TV the night before an exam decreases a student's exam grade by 8 points.

   B. Watching one additional hour of TV the night before an exam decreases a student's exam grade by 100 points.

   C. Each additional hour of TV watching increases a student's grade by 8 points.

   D. A student would need to watch a minimum of 100 hours of TV to have his or her grade affected.

9. Last year, when Mr. Smith bought a new washer/dryer combo, he collected pairs of data in a scatter plot to help him understand how the value of his appliance would change over time. He then created and graphed the linear model shown below.

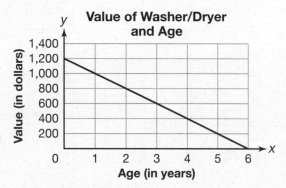

A. Identify the *x*- and *y*-intercepts. What does each tell you about the problem situation?

_____

_____

_____

B. Identify the slope of the graph. Use the slope to describe the association between the age of the washer/dryer and its value.

_____

_____

Patterns in Data

Common Core State Standard:
8.SP.4

Getting the Idea

The **frequency** of a piece of data is the number of times it appears in a data set. A **frequency distribution** is a way of grouping data so that meaningful patterns can be found. A frequency distribution table is used to show the total for each category or group.

Example 1

A sporting goods store recorded the number of tents sold each week in February and used the data to make a frequency table.

Tents Sold

Week	Tallies	Frequency
1	⁃⁃⁃⁃	5
2	⁃⁃⁃ ⁃⁃⁃ ⁃⁃	12
3	⁃⁃⁃ ⁃⁃	7
4	⁃⁃⁃ ⁃⁃⁃	10
Total		**34**

What percentage of all the tents sold in February were sold during the first two weeks of the month?

Strategy **Use the frequency distribution table.**

Step 1 Look at the frequency column for Weeks 1 and 2.

5 tents were sold during Week 1.

12 tents were sold during Week 2.

So, 5 + 12, or 17, tents were sold during that two-week period.

Step 2 Find the percentage.

The table shows that a total of 34 tents were sold in February.

$\frac{17}{34} = 0.5 = 0.5 \times 100\% = 50\%$

Solution **Exactly 50% of the tents sold in February were sold during the first two weeks of the month.**

In Example 1, you calculated the percentage of tents that were sold during a certain time period. Sometimes, it is helpful to show percentages in a frequency table by including a column for **relative frequency**. The relative frequency of a given category is found by dividing the frequency of that category by the sum of all the frequencies.

Example 2

The number of computers owned by the family of each student in Ms. Fontana's class is shown below.

0, 2, 1, 3, 2, 4, 0, 2, 2, 3, 1, 2, 3, 3, 2, 0, 1, 4, 2, 3, 5, 2, 1, 5, 0

Here is how that information looks in a partially-completed frequency distribution table.

Computers Owned by Students' Families

Computers Owned	Frequency	Relative Frequency
0	4	16%
1	4	16%
2	8	32%
3	5	
4	2	
5	2	
Total	25	100%

Complete the last column of the table. Check that the relative frequencies total 100%.

Strategy   **Find the relative frequencies that are missing from the table.**

Step 1   Find the relative frequency for the "3 computers" row.

Five of 25 students' families have 3 computers.

$\frac{5}{25} = 0.2 = 0.2 \cdot 100\% = 20\%$

Note: Relative frequency can be shown as either a decimal or a percent.

Step 2   Find the relative frequency for the other two rows.

Since both rows show frequencies of 2, their relative frequencies will be the same.

$\frac{2}{25} = 0.08 = 0.08 \cdot 100\% = 8\%$

Step 3   Be sure that all the relative frequencies add up to 100%.

$16\% + 16\% + 32\% + 20\% + 8\% + 8\% = 100\%$ ✓

Solution   **The row for "3 computers" should show 20% as the relative frequency. The rows for "4 computers" and "5 computers" should each show 8% as the relative frequency.**

In Examples 1 and 2, we compared the frequencies of data collected about one variable. Sometimes, you may want to look at two variables at the same time and determine if there is a relationship between them. One way to do that is to create a **two-way table**. You can enter either frequency counts or relative frequencies in the cells of the table.

Example 3

Ten students in a class were asked two questions. They were asked to tell if they do chores at home or not. They were then asked if they receive an allowance or not. The results are shown below.

Student Survey

Student	Abby	Bella	Chris	Deb	Erin	Frank	Gus	Hal	Isadore	John
Chores	Yes	Yes	No	No	No	Yes	Yes	Yes	No	Yes
Allowance	Yes	Yes	No	No	No	Yes	Yes	No	Yes	No

Create a two-way table to show the frequency counts for these data.

Strategy **Determine how the table will look. Then fill in the frequencies.**

Step 1 Determine how the table will look.

Include a column for "allowance" and a column for "no allowance" along the top.

Include a row for "chores" and a row for "no chores" along the left side.

Step 2 Decide how to fill in the first row of cells.

Four students (Abby, Bella, Frank, and Gus) do chores and get an allowance.

The first cell shows "chores" and "allowance." Record 4 in the first cell.

Two students (Hal and John) do chores and get no allowance. Record 2 in the second cell.

Find and record the total for that row: 4 + 2 = 6.

	Allowance	No Allowance	Total
Chores	4	2	6
No Chores			
Total			

Step 3 Complete the second row in the table. Then add the columns and record those totals.

	Allowance	No Allowance	Total
Chores	4	2	6
No Chores	1	3	4
Total	5	5	10

Solution **The two-way table in Step 3 organizes the data.**

You can also use a two-way table to display relative frequencies instead of actual frequencies. You can show the relative frequencies for each row or for each column.

Example 4

Look back at the two-way table in Example 3. Can you conclude that students who get an allowance are more likely to do chores than students who do not? Find the relative frequencies for the columns in the table and see.

	Allowance	No Allowance	Total
Chores	4	2	6
No Chores	1	3	4
Total	5	5	10

Strategy   **Recreate the two-way table, showing the relative frequencies for the columns.**

Step 1   Find the relative frequencies.

Of all students who get an allowance, $\frac{4}{5}$, or 80%, also do chores, while $\frac{1}{5}$, or 20%, do not.

Of all students who do not get an allowance, $\frac{2}{5}$, or 40%, also do chores, while $\frac{3}{5}$, or 60%, do not.

	Allowance	No Allowance	Total
Chores	80%	40%	60%
No Chores	20%	60%	40%
Total	100%	100%	100%

Step 2   What conclusions can you draw?

The data do show that students who do chores are more likely to get an allowance than students who do not do chores, because 80% of the students who get an allowance also do chores.

Solution   **The data indicate a positive association between getting an allowance and doing chores.**

Coached Example

The P.E. teachers want to offer students a choice of several electives: yoga, flag football, or ultimate Frisbee. They wanted to see if the boys and girls had different first choices. So, they surveyed 200 students and made a two-way table that showed the relative frequencies for the rows, as shown below.

	Yoga	Flag Football	Frisbee	Total
Boys	0.05	0.65	0.30	1.00
Girls	0.60	0.05	0.35	1.00
Total	0.325	0.35	0.325	1.00

Is there any P.E. elective that seems to be about as popular among both boys and girls?

Since the table shows the relative frequencies for each row, you can look at the relationship between gender and P.E. elective preferences.

The probability that a girl will prefer yoga is _____, while the probability that a boy will prefer yoga is _____. So, yoga is more popular among the _____.

Similarly, _____ are more likely to prefer flag football.

The probability that a boy will prefer Frisbee is _____, and the probability that a girl will prefer Frisbee is _____.

What does that tell you about that elective choice?

Of all the electives, _____ seems to be similarly popular among both boys and girls.

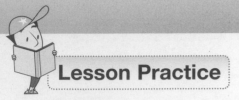

Lesson Practice

Choose the correct answer.

1. Eight students in a class were asked two questions. They were asked if they play on a sports team and if they are learning to play a musical instrument. Each row below represents a student's answer.

Team?	no	no	yes	yes	yes	no	no	yes
Instrument?	yes	no	no	no	yes	yes	yes	yes

Which two-way table best displays these data?

A.

	Instrument	No Instrument	Total
Team	4	0	4
No Team	4	0	4
Total	4	0	8

B.

	Instrument	No Instrument	Total
Team	2	2	4
No Team	3	1	4
Total	5	3	8

C.

	Instrument	No Instrument	Total
Team	1	3	4
No Team	2	2	4
Total	3	5	8

D.

	Instrument	No Instrument	Total
Team	2	2	4
No Team	1	3	4
Total	3	5	8

Use the table below for questions 2–4.

A survey of students in a class explored the relationship between gender and vegetarianism.

	Vegetarian	Non-vegetarian	Total
Boys	3	12	15
Girls	6	9	15
Total	9	21	30

2. If a girl is chosen from the class at random, what is the probability that she is a vegetarian?

   A. 6%

   B. 20%

   C. 40%

   D. 60%

3. If a boy is chosen from the class at random, what is the probability that he is **not** a vegetarian?

   A. 80%

   B. 62%

   C. 50%

   D. 12%

4. Which is a reasonable conclusion to draw from these data?

   A. There are more vegetarians in the class than non-vegetarians.

   B. Among the boys, there are more vegetarians than non-vegetarians.

   C. Among the girls, there are more vegetarians than non-vegetarians.

   D. The probability that a girl chosen from the class is a vegetarian is greater than the probability that a boy is a vegetarian.

5. Cathy wanted to see if there was a relationship between students' grade levels and school club participation. She made this two-way table to show her results.

	One or More Clubs	Not in Clubs	Total
Grade 6	6	44	50
Grade 7	23	27	50
Grade 8	40	10	50
Total	69	81	150

A. Find the relative frequencies for the table above. Record those frequencies below.

	One or More Clubs	Not in Clubs	Total
Grade 6			
Grade 7			
Grade 8			
Total			

B. Draw and state two conclusions about the relationship between a student's grade level and the likelihood that he or she will participate in school clubs.

_____

_____

_____

_____

Domain 5: Cumulative Assessment for Lessons 33–36

Use the scatter plot for questions 1 and 2.

The scatter plot below compares the number of pages that were in a fiction book that students read to the number of days it took to finish it.

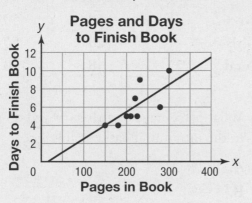

1. What type of association, if any, is shown by the scatter plot?

   A. negative association

   B. no association

   C. nonlinear association

   D. positive association

2. Which statement is true about the line of best fit drawn above?

   A. The line shows the correct association, but it does not come close to most points.

   B. The line comes close to nearly all points, so it is a very good model for the data.

   C. The scatter plot shows no association, so a line should not be used to model the data.

   D. The data do not resemble a straight line, so a nonlinear model would be better for these data.

Use the scatter plot for questions 3 and 4.

The scatter plot below compares the number of hot pretzels sold at a concession stand to the number of beverages sold.

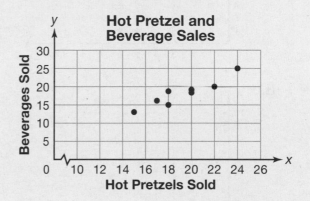

3. If the data above contain an outlier, which coordinates best represent it?

   A. (15, 13)

   B. (20, 19)

   C. (24, 5)

   D. There is no outlier for these data.

4. Which best describes the association shown by the scatter plot?

   A. positive, linear association

   B. negative, linear association

   C. nonlinear association

   D. no association

5. Which scatter plot shows a negative, linear association for the data?

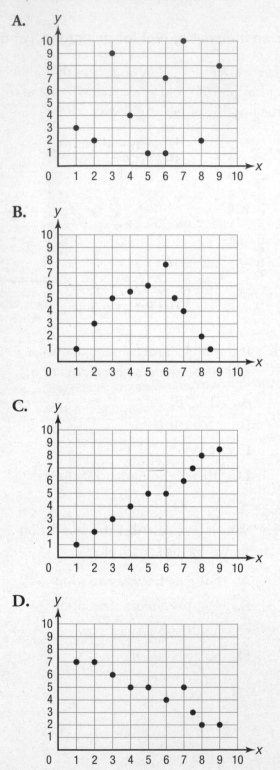

A. y

B. y

C. y

D. y

A survey of randomly-selected eighth-grade students explored the relationship between gender and participation in a school intramural basketball league.

	Play in League	Do Not Play in League	Total
Boys	31	19	50
Girls	21	29	50
Total	52	48	100

6. What percentage of girls surveyed do play in the intramural basketball league?

A. 72%

B. 58%

C. 42%

D. 21%

7. Which is **not** a reasonable interpretation of the data?

A. Ignoring gender, roughly half of the students surveyed play in the league.

B. More girls surveyed play in the league than boys do.

C. Approximately 60% of the girls surveyed do not play in the league.

D. Approximately 38% of the boys surveyed do not play in the league.

Use the information for questions 8 and 9.

A restaurant manager collected data and compared the prices of different sandwiches to the number sold. She found that the linear model $y = 59 - 4x$ could be used to predict y, the number of each type of sandwich sold, if the sandwich cost x dollars.

8. What association is shown by the linear model?

   A. With each additional dollar that is added to the price of a sandwich, the number sold decreases by 4.

   B. With each additional dollar that is added to the price of a sandwich, the number sold decreases by 59.

   C. With each additional dollar that is added to the price of a sandwich, the number sold increases by 4.

   D. With each additional dollar that is added to the price of a sandwich, the number sold increases by 59.

9. Based on the linear model, how many sandwiches should the restaurant manager expect to sell if the sandwich costs $8.50?

   _____

10. Eight students in a class were asked two questions. They were asked if they eat breakfast every morning and if they exercise every day. Each column below represents a student's answer.

Breakfast?	yes	yes	yes	yes	no	no	no	yes
Exercise?	yes	yes	no	no	no	yes	no	yes

   A. Complete the two-way table below to show these results.

	Exercise	No Exercise	Total
Breakfast			
No Breakfast			
Total			

   B. Create a second two-way table that shows relative frequencies calculated for the columns from the table in Part A.

	Exercise	No Exercise	Total
Breakfast			
No Breakfast			
Total			

Glossary

AA (Angle-Angle) similarity theorem states that two triangles are similar if two angles of one triangle are congruent to two angles of the other triangle (Lesson 26)

absolute value the distance of a number from zero on the number line (Lesson 30)

addition property of equality states that if you add the same number to both sides of an equation, the equation continues to be true (Lesson 9)

alternate exterior angles angles that lie outside the parallel lines and are on opposite sides of the transversal (Lesson 28)

alternate interior angles angles that lie inside the parallel lines and are on opposite sides of the transversal (Lesson 28)

base in a power, the number that is used as a factor the number of times indicated by the exponent (Lesson 5)

coefficient a number that is multiplied by a variable in an algebraic term (Lesson 17)

coincident lines lines that lie on top of one another; coincident lines share all points in common (Lesson 15)

congruent same shape and same size; polygons are congruent if all corresponding angles and all corresponding sides are equal (Lesson 24)

constant of proportionality the nonzero constant m in the function $y = mx$ (Lesson 14)

corresponding angles angles that lie on the same side of the transversal and on the same side of the parallel lines (Lesson 28)

cube root one of three equal factors of a number (Lesson 6)

dependent variable a variable that provides the output values of a function (Lesson 20)

diameter the widest part of a circle; twice the length of the radius (Lesson 32)

dilation a transformation that changes the size of a figure according to a scale factor (Lesson 25)

direct proportion a linear equation that can be expressed in the form $y = mx$, where $m \neq 0$ (Lesson 14)

distributive property states that the product of a number and an indicated sum is the same as the sum of the products of the number and each addend of the indicated sum (Lesson 9)

division property of equality states that if you divide both sides of an equation by the same number, the equation continues to be true (Lesson 9)

domain the input values in a function (Lesson 19)

enlargement a dilation that stretches a figure (Lesson 25)

equation a mathematical sentence with an equal (=) sign (Lesson 9)

exponent in a power, the number that indicates how many times the base is used as a factor (Lesson 5)

expression a mathematical arrangement of numbers and/or variables connected by one or more operations (Lesson 4)

exterior angle an angle formed by a side of a polygon and an extension of an adjacent side (Lesson 27)

finite decimal a decimal with digits that terminate (Lesson 1)

frequency the number of times that a piece of data appears in a set (Lesson 36)

frequency distribution a list of values in a sample, showing the number of times each value occurs (Lesson 36)

function a set of ordered pairs in which each input value corresponds to exactly one output value (Lesson 19)

image the figure resulting from a transformation (Lesson 24)

independent variable a variable that provides the input values of a function (Lesson 20)

integers the set of numbers consisting of the natural numbers (1, 2, 3, …), their opposites (−1, −2, −3, …), and zero (Lesson 1)

interior angle an angle that is on the inside of a polygon and has its vertex formed by two sides of the polygon (Lesson 27)

intersecting lines lines that cross one another (Lesson 15)

inverse operations operations that undo one another (Lesson 9)

irrational number any number that is not rational; an irrational number can be expressed as a non-repeating, non-terminating decimal (Lesson 2)

isolate the variable manipulate a one-variable linear equation so that the variable is by itself on one side of the equation (Lesson 9)

like terms terms that have the same variable raised to the same power, or terms that have the same radical expression (Lesson 9)

line of best fit the line that most closely represents the relationship between the two variables in a scatter plot; also called a trend line (Lesson 34)

linear equation an equation that represents a straight line (Lesson 9)

linear function a function that has a constant rate of change and whose graph is a straight line (Lesson 20)

multiplication property of equality states that if you multiply both sides of an equation by the same number, the equation continues to be true (Lesson 9)

nonlinear function a function that does not have a constant rate of change and whose graph is not a straight line (Lesson 20)

origin the point (0, 0) on a coordinate plane where the x-axis and y-axis intersect (Lesson 12)

outlier a data point with values that are significantly different from the other data points in the set (Lesson 33)

parallel lines lines that lie in the same plane and never intersect (Lessons 15, 28)

perfect cube a number whose cube root is a whole number (Lesson 6)

perfect square a number whose square root is a whole number (Lesson 1)

point-slope form a linear equation of a non-vertical line that passes through a point (x_1, y_1) and has slope m. It is written in the form $y - y_1 = m(x - x_1)$. (Lesson 12)

power a number raised to a given exponent (Lesson 5)

power of a power (property of exponents) to raise a power to a power, multiply the exponents (Lesson 5)

power of a product (property of exponents) to find a power of a product, find the power of each factor and multiply (Lesson 5)

power of a quotient (property of exponents) to raise a quotient to a power, raise both the numerator and denominator to that power (Lesson 5)

power of zero (property of exponents) any nonzero number raised to the power of zero is 1 (Lesson 5)

product of powers (property of exponents) to multiply two numbers with the same base, add the exponents (Lesson 5)

proportion an equation stating that two ratios are equal in value (Lesson 13)

Pythagorean theorem states that the sum of the squares of the lengths of the legs in a right triangle is equal to the square of the hypotenuse (Lesson 29)

quotient of powers (property of exponents) to divide two numbers with the same base, add the exponents (Lesson 5)

radical the symbol $\sqrt{}$ which indicates the principal square root of a number (Lesson 6)

radicand the number under a radical sign (Lesson 6)

radius a line segment from the center of a circle to any point on the circle; half the length of the diameter (Lesson 32)

range the output values in a function (Lesson 19)

rate a comparison of two quantities that uses different units of measure (Lesson 13)

rate of change a ratio that compares two quantities (Lessons 11, 20)

ratio a comparison of two numbers (Lesson 11)

rational number any number that can be expressed as the ratio of two integers, excluding division by zero (Lesson 1)

real number any number from the set that includes rational and irrational numbers (Lesson 2)

reciprocals two numbers whose product is 1 (Lesson 5)

reduction a dilation that shrinks a figure (Lesson 25)

reflection a flip of a figure over a point or a line (Lesson 24)

relation a set of ordered pairs (Lesson 19)

relative frequency the frequency of a particular category divided by the sum of all the frequencies, expressed as a percent or a decimal (Lesson 36)

repeating decimal a decimal with one or more digits that repeat forever (Lesson 1)

rigid transformation a movement of a figure in a plane; its size and shape do not change (Lesson 24)

rotation a turn of a figure about a point (Lesson 24)

rule a procedure for generating a pattern or function (Lesson 19)

SAS (Side-Angle-Side) similarity theorem states that two triangles are similar if an angle of one triangle is congruent to an angle of the other triangle and if the lengths of the sides that include the angles are proportional (Lesson 26)

scale factor the ratio of the lengths of corresponding sides of two similar figures (Lesson 25)

scatter plot a graph of paired data in which the data values are plotted as points in (x, y) format (Lesson 33)

scientific notation a way to abbreviate very large or very small numbers using powers of 10 (Lesson 7)

similar triangles triangles that have the same shape, but not necessarily the same size (Lessons 11, 26)

slope a ratio that compares the change in the y-coordinates of a graph to the change in the x-coordinates (Lesson 11)

slope-intercept form a representation of a linear equation as $y = mx + b$, where m represents the slope and b represents the y-intercept (Lesson 12)

square root one of the two equal factors of a number (Lesson 1)

straight angle an angle that measures 180° (Lesson 27)

subtraction property of equality states that if you subtract the same number from both sides of an equation, the equation continues to be true (Lesson 9)

supplementary angles whose measures add up to 180° (Lesson 27)

system of linear equations a set of two or more linear equations that use the same variables (Lesson 16)

translation a slide of a figure to a new location (Lesson 24)

transversal a line that intersects two or more lines at different points (Lesson 28)

trend line the line that most closely represents the relationship between the two variables in a scatter plot; also called a line of best fit (Lesson 34)

two-way table a table that displays two different variables for the same input at the same time (Lesson 36)

unit price a unit rate in which the numerator is an amount of money (Lesson 13)

unit rate a rate that, when expressed as a fraction, has a 1 in the denominator (Lesson 13)

variable a letter or symbol that represents a number or set of numbers (Lesson 9)

vertical line test a method to determine if a relation is a function; a relation is a function if no vertical line passes through its graph in more than one point (Lesson 19)

volume the number of cubic units that fit inside a three-dimensional figure (Lesson 32)

whole number a number from the set that includes the counting numbers (1, 2, 3, ...) and zero (Lesson 1)

x-intercept the point or points at which a graph crosses the x-axis (Lesson 35)

y-intercept the point or points at which a graph crosses the y-axis (Lesson 12)

Summative Assessment:
Domains 1–5

Name: _____

Session 1

1. What is the value of $\sqrt[3]{27}$?

   A. 1

   B. 2

   C. 3

   D. 9

2. Which symbol goes in the blank to make this sentence true?

   $$\sqrt{15} \bigcirc \pi$$

   A. $>$

   B. $<$

   C. $=$

   D. $+$

3. What is the value of x in $x^2 = 14$?

   A. 1.4

   B. 7

   C. $\sqrt{14}$

   D. $\sqrt[3]{14}$

4. Which graph does **not** represent a function?

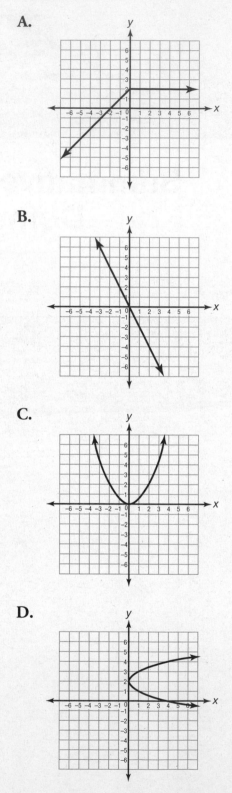

   A.

   B.

   C.

   D.

5. Triangle *PQR* and triangle *STV* are shown on the coordinate grid below.

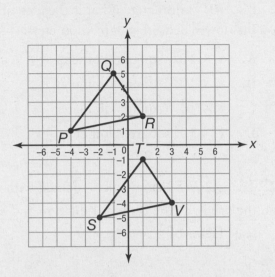

If △*PQR* is translated 6 units down and 2 units to the right, which will be true?

A. Angle *P* will move onto congruent ∠*S*.

B. Angle *P* will move onto congruent ∠*T*.

C. Angle *R* will move onto congruent ∠*S*.

D. Angle *R* will move onto congruent ∠*T*.

6. The distance across a lake cannot be directly measured. A land surveyor takes some other measurements and uses them to find *d*, the distance across the lake. What is the value of *d*?

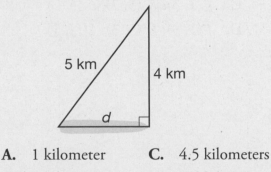

A. 1 kilometer   C. 4.5 kilometers

B. 3 kilometers   D. 6.4 kilometers

Use the scatter plot for questions 7 and 8.

The scatter plot below shows the daily high temperature and the number of cups of hot cocoa sold by a cafe over a 3-week period.

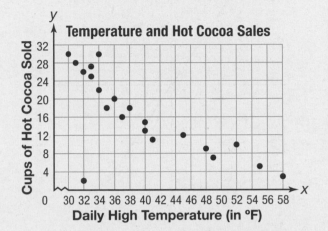

7. If the data contain an outlier, which coordinates best represent it?

A. (30, 30)

B. (32, 2)

C. (58, 3)

D. There is no outlier for these data.

8. Which best describes the association shown by the scatter plot?

A. negative, linear association

B. positive, linear association

C. nonlinear association

D. no association

9. What is the equation for the line graphed below?

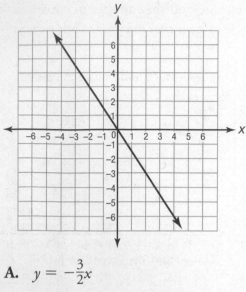

A. $y = -\frac{3}{2}x$

B. $y = -\frac{2}{3}x$

C. $y = \frac{2}{3}x$

D. $y = \frac{3}{2}x$

10. Which formula below describes a linear function?

A. area of a square with sides s units long: $A = s^2$

B. surface area of a sphere with radius r units long: $A = 4\pi r^2$

C. perimeter of a square with sides s units long: $P = 4s$

D. volume of a cube with edges s units long: $V = s^3$

11. A rectangular flower bed has a length of 0.005 kilometer and a width of 8×10^{-3} kilometer. What is the area of the flower bed, in square kilometers?

A. 4×10^{-7} square kilometer

B. 4×10^{-6} square kilometer

C. 4×10^{-5} square kilometer

D. 4×10^{6} square kilometers

12. Triangle *JKL* below will be dilated with the origin as the center of dilation and a scale factor of $\frac{1}{4}$.

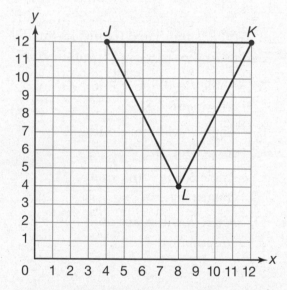

What will be the coordinates of the vertices of the dilated image, $\triangle J'K'L'$?

A. $J'(16, 48)$, $K'(48, 48)$, $L'(32, 16)$

B. $J'(1, 12)$, $K'(3, 12)$, $L'(2, 4)$

C. $J'(1, 3)$, $K'(3, 3)$, $L'(4, 2)$

D. $J'(1, 3)$, $K'(3, 3)$, $L'(2, 1)$

13. What is the value of $4^5 \div 4^2$?

    A. 81

    B. 64

    C. 16

    D. 12

14. Last semester, Mr. Oppenheim collected data to determine the association between the number of hours that students played video games the night before a test and students' test scores. The equation $y = 100 - 10x$ was a good linear model for determining y, a student's score on the test, after playing video games for x hours the night before. What association is shown by the linear model?

    A. Each additional hour of video game playing decreases a student's test score by 10 points.

    B. Each additional hour of video game playing increases a student's test score by 10 points.

    C. The number of hours spent playing video games has no effect on students' test scores.

    D. A student would need to play video games for 100 hours in order to have his or her test score affected.

15. A fully inflated beach ball has a diameter of 16 inches. Which is closest to the amount of air inside the ball?

    A. 268 in.^{3}

    B. 1,072 in.^{3}

    C. 2,144 in.^{3}

    D. 17.149 in.^{3}

16. Compare the rates of change for the two functions described below.

Function 1 is represented by the equation $y = 3x - 6$.

The table below represents Function 2.

x	-2	0	2	4	6
y	-14	-6	2	10	18

Which statement about the two functions is true?

    A. Both functions have the same rate of change.

    B. Function 1 has a greater rate of change than Function 2.

    C. Function 2 has a greater rate of change than Function 1, and both functions have different y-intercepts.

    D. Function 2 has a greater rate of change than Function 1, but both functions have the same y-intercept.

17. Which best describes the solution for this equation?

$$8p - 12 = 4(2p - 12)$$

    A. $p = \frac{3}{2}$

    B. $p = 6$

    C. no solution

    D. infinitely many solutions

18. Trapezoid *KLMN* and trapezoid *WXYZ* are shown on the coordinate grid below.

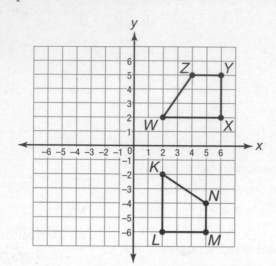

If trapezoid *KLMN* is rotated 90° counterclockwise about the origin, onto which congruent line segment will $\overline{KL}$ move?

A. $\overline{WX}$

B. $\overline{XY}$

C. $\overline{YZ}$

D. $\overline{ZW}$

19. Which point on the number line best represents $\sqrt{7}$?

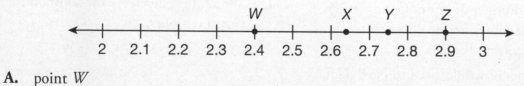

A. point *W*

B. point *X*

C. point *Y*

D. point *Z*

20. Solve the system of linear equations.

$$3x - 4y = 4$$
$$3x - 4y = -8$$

  A. $(0, -1)$

  B. $(4, -1)$

  C. no solution

  D. infinitely many solutions

21. Which best describes the solution for this equation?

$$-2(x + 3) = -2x - 6$$

  A. $x = -6$

  B. $x = 6$

  C. no solution

  D. infinitely many solutions

22. The population of New York is approximately 2×10^7. The population of New Jersey is approximately 9×10^6. Which statement accurately compares the populations of New York and New Jersey?

  A. The population of New Jersey is about 45 times greater than the population of New York.

  B. The population of New York is about 45 times greater than the population of New Jersey.

  C. The population of New Jersey is more than 2 times greater than the population of New York.

  D. The population of New York is more than 2 times greater than the population of New Jersey.

23. Which statement is true of the interval from $x = 1$ to $x = 4$?

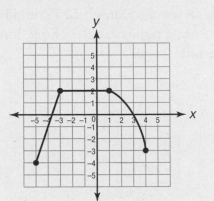

  A. That piece of the graph is nonlinear.

  B. That piece of the graph is linear.

  C. That piece of the graph is increasing.

  D. That piece of the graph is constant.

24. Which sequence could be used to show that $\triangle QRS$ is congruent to $\triangle WXY$?

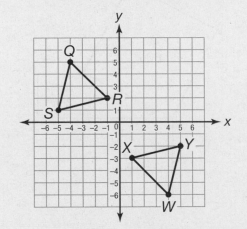

  A. 90° clockwise rotation of $\triangle QRS$ around the origin followed by a translation of 1 unit down

  B. 180° rotation of $\triangle QRS$ around the origin followed by a translation of 1 unit down

  C. 180° rotation of $\triangle QRS$ around the origin followed by a translation of 1 unit to the left

  D. No sequence of transformations could be used because $\triangle QRS$ and $\triangle WXY$ are not congruent.

25. Elizabeth needs to identify a right triangle. When joined at the vertices, which set of squares below could be used to form a right triangle? Note: drawings are not to scale.

A.

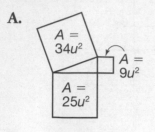

B.

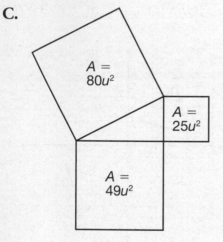

C.

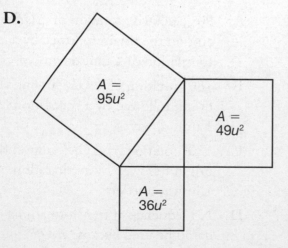

D.

26. A chef teaches private cooking classes. She charges a set fee for each class plus an additional rate for each hour she teaches, as shown by the graph below.

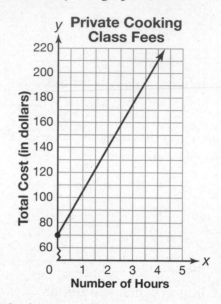

Which statement is true about this graph?

A. The y-intercept shows that the set fee is $0.

B. The y-intercept shows that the set fee is $70.

C. The slope shows that the hourly rate is $70 per hour.

D. The slope shows that the hourly rate is $210 per hour.

27. Elena took the angles of △*DEF* and rearranged them to show that, together, they can form a straight line. Which is true of the measures of these angles?

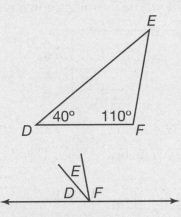

A. Angle *E* must measure 40°.

B. Angle *E* must measure 110°.

C. The sum of the measures of angles *D*, *E*, and *F* must measure 90°.

D. The sum of the measures of angles *D*, *E*, and *F* must measure 180°.

28. What is the equation of this line?

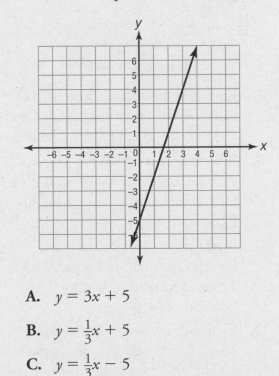

A. $y = 3x + 5$

B. $y = \frac{1}{3}x + 5$

C. $y = \frac{1}{3}x - 5$

D. $y = 3x - 5$

29. Which sequence could be used to show that rectangle *ABCD* is similar to rectangle *WXYZ*?

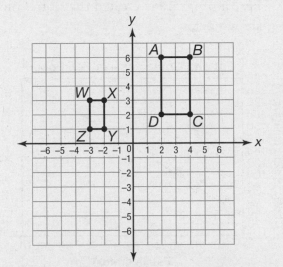

A. dilation of *ABCD* by a scale factor of $\frac{1}{2}$ followed by a translation of 1 unit down

B. dilation of *ABCD* by a scale factor of $\frac{1}{2}$ followed by a translation of 4 units to the left

C. dilation of *ABCD* by a scale factor of $\frac{1}{4}$ followed by a translation of 1 unit down

D. dilation of *ABCD* by a scale factor of $\frac{1}{4}$ followed by a translation of 4 units to the left

30. Solve the system of linear equations.

$$2x + y = 3$$
$$6x - 2y = 14$$

A. $(1, -4)$

B. $(1, 1)$

C. $(2, -1)$

D. $(2, 1)$

31. Tim compared the number of grams of fat to the number of milligrams of cholesterol in his 10 favorite foods for a health project. He created the scatter plot shown below.

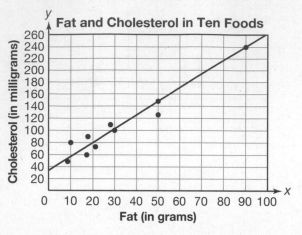

Which statement is true about the line of best fit?

A. The line comes close to nearly all points, so it is a very good model for the data.

B. The line shows the correct association, but does not come close to most points.

C. The scatter plot shows no association, so a line should not be used to model the data.

D. The data do not resemble a straight line, so a nonlinear model would be better for these data.

32. The size of the rectangular display for a computer monitor is given as *d*, the distance between two opposite screen corners. What is *d*, the best estimate of the size of the monitor shown below?

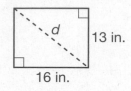

A. 9.3 inches   **C.** 29 inches

B. 20.6 inches   **D.** 213 inches

33. The graph shows the distance that Brad traveled to and from his home on a motorized scooter. Which best describes what is shown by the graph?

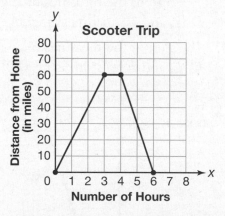

A. Brad rode up a hill for 3 hours, then along the flat top of the hill for 1 hour, and then down the hill for 2 hours.

B. Brad rode down a hill for 3 hours, then along the flat top of the hill for 1 hour, and then up the hill for 2 hours.

C. Brad rode away from his home for 3 hours, then took a break for 1 hour, and then rode toward his home for 2 hours.

D. Brad rode toward his home for 3 hours, then stayed at his home for 1 hour, and then rode away from his home for 2 hours.

34. Parallelogram *ABCD* and parallelogram *FGHI* are shown on the coordinate system below.

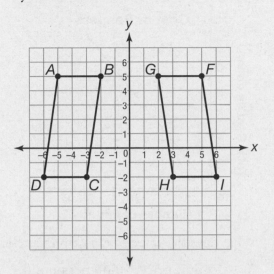

If parallelogram *ABCD* is reflected over the *y*-axis so it completely covers parallelogram *FGHI*, which will be true?

A. Parallel line segments *AB* and *DC* will be moved onto parallel line segments *GF* and *HI*.

B. Parallel line segments *AB* and *DC* will be moved onto parallel line segments *GH* and *FI*.

C. Parallel line segments *AD* and *BC* will be moved onto parallel line segments *GF* and *FI*.

D. Parallel line segments *AD* and *BC* will be moved onto parallel line segments *GF* and *HI*.

35. Which is equivalent to $2^4 \cdot 2^{-6}$?

A. -4

B. $-\dfrac{1}{4}$

C. $\dfrac{1}{4}$

D. 4

36. During a sale at an online shoe store, each pair of sandals costs $15. The store charges a shipping and handling fee of $5, no matter how many pairs of sandals a customer orders. Which equation best represents *y*, the total cost in dollars of buying *x* pairs of sandals from this online store?

A. $15x + 5 = y$

B. $15x + 5x = y$

C. $15 + 5x = y$

D. $15 + 5 + x = y$

37. William is writing a report on dinosaurs. He learns that the plant-eating dinosaur *Argentinosaurus* could grow to a weight of about 2×10^5 pounds. The meat-eating dinosaur *Allosaurus* could grow to a weight of about 9×10^3 pounds. Which is an accurate comparison of the weights of these two dinosaurs?

A. *Argentinosaurus* weighed about 200 times as much as *Allosaurus*.

B. *Allosaurus* weighed about 200 times as much as *Argentinosaurus*.

C. *Argentinosaurus* had a weight that was less than 20 times that of *Allosaurus*.

D. *Argentinosaurus* had a weight that was more than 20 times that of *Allosaurus*.

38. Which is a rational number that can be written as a decimal in which one or more non-zero digits repeat?

A. 23.5% **C.** $\sqrt{3}$

B. $\dfrac{5}{6}$ **D.** π

39. A cone-shaped paper cup has a diameter of 6 centimeters and a height of 9 centimeters.

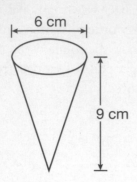

Approximately how many cubic centimeters of water can the cup hold?

A. 85 cm³   **C.** 339 cm³

B. 254 cm³   **D.** 1,017 cm³

40. Which equation does **not** represent a linear function?

A. $y = -\frac{1}{2}x$ **C.** $y = -2x - 2$

B. $y = -x - 2$ **D.** $y = -x^2$

41. $\triangle RST$ is similar to $\triangle WXY$ because two pairs of corresponding angles are congruent. Which is **not** true?

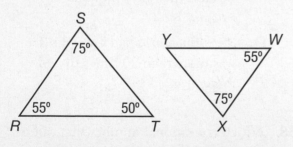

A. The sum of the angle measures of $\triangle WXY$ is 180°, so $m\angle Y = 50°$.

B. $m\angle R = m\angle W = 55°$

C. $m\angle S = m\angle X = 75°$

D. $m\angle T = m\angle W = 55°$

42. Leo is buying peppers from a farm stand. The graph below shows the cost.

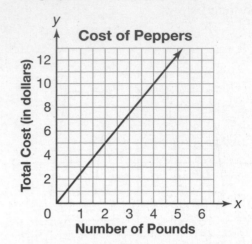

What does the slope of the graph represent?

A. the unit price, $0.40 per pound of peppers

B. the unit price, $2.50 per pound of peppers

C. the number of pounds Leo bought, 4 pounds of peppers

D. the total amount Leo spent, $10.00

43. Triangle ABC below will be rotated 90° clockwise about the origin.

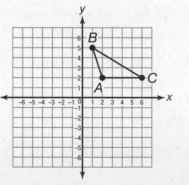

What will be the coordinates of the vertices of the dilated image, $\triangle A'B'C'$?

A. $A'(2, -2)$, $B'(5, -1)$, $C'(2, -6)$

B. $A'(2, 2)$, $B'(-5, 1)$, $C'(-6, -2)$

C. $A'(-2, 2)$, $B'(-5, 1)$, $C'(-2, 6)$

D. $A'(-2, -2)$, $B'(-1, -5)$, $C'(-6, -2)$

44. Compare the rates of change for the two linear functions represented below.

Function 1

x	y
−6	−2
−3	0
0	2
3	4
6	6

Function 2

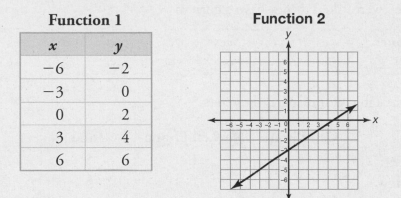

Which statement about the two functions is true?

A. Both functions have the same rate of change.

B. Function 1 has a greater rate of change than Function 2.

C. Function 2 has a greater rate of change than Function 1, and both functions have different y-intercepts.

D. Function 2 has a greater rate of change than Function 1, but both functions have the same y-intercept.

45. Calliope and Matt both walk dogs to earn extra money, and each charges an hourly rate. The equation $y = 8.25x$ shows how to calculate y, the total charge in dollars, for Calliope to walk a dog for x hours. The table below shows the information for Matt.

Matt's Charges

x	2	4	6	8
y	17	34	51	68

Which statement is true?

A. Calliope's hourly rate is $0.25 cheaper than Matt's.

B. Matt's hourly rate is $0.25 cheaper than Calliope's.

C. Calliope's hourly rate is $6.25 cheaper than Matt's.

D. Calliope and Matt charge the same hourly rate.

46. Eight students in a class were asked two questions. They were asked if they have a personal cell phone and they were asked if they have a curfew.

Cell phone?	yes	yes	yes	yes	no	yes	yes	no
Curfew?	yes	no	yes	yes	no	yes	yes	no

Which two-way table best displays these data?

A.

	Cell Phone	No Cell Phone	Total
Curfew	4	1	5
No Curfew	2	1	3
Total	6	4	8

B.

	Cell Phone	No Cell Phone	Total
Curfew	5	0	5
No Curfew	2	1	3
Total	7	1	8

C.

	Cell Phone	No Cell Phone	Total
Curfew	5	0	5
No Curfew	1	2	3
Total	6	2	8

D.

	Cell Phone	No Cell Phone	Total
Curfew	4	0	4
No Curfew	4	0	4
Total	8	0	8

47. Which best describes the graphs of the line that passes through (0, 2) and (6, 4) and the line that passes through (2, 1) and (5, 7)?

A. The lines are coincident.

B. The lines intersect in exactly one point.

C. The lines are parallel.

D. The lines lie in the same plane, but never intersect.

48. Which best describes the solution for this equation?

$$0.5(4x + 3) = 5x - 2.5$$

- **A.** $x = 0.75$
- **B.** $x = 1.\overline{3}$
- **C.** $x = 4$
- **D.** $x = 12$

49. On the map of Franklinton below, the museum is at the origin (0, 0) and each unit represents 1 kilometer.

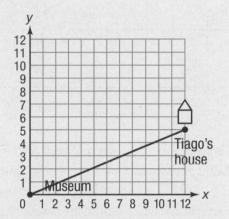

Tiago lives 5 kilometers north and 12 kilometers east of the museum. If he rides his bike directly from his house to the museum one day, how far will he ride his bike?

- **A.** 7 kilometers
- **B.** 13 kilometers
- **C.** 17 kilometers
- **D.** 84.5 kilometers

50. The triangle below has angles measuring a, b, and c degrees and an exterior angle measuring $x°$. Carlos drew two parallel lines and used what he knows about angles formed when parallel lines are cut by a transversal to find two other angle measures.

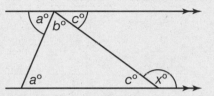

Using the information above, which expression is equivalent to $x°$?

- **A.** $180° - a°$
- **C.** $a° + b°$
- **B.** $180° - b°$
- **D.** $a° + c°$

51. Which sequence could be used to show that $\triangle DEF$ is similar to $\triangle GHJ$?

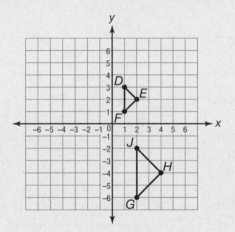

- **A.** dilation of $\triangle DEF$ by a scale factor of 2 followed by a reflection across the x-axis
- **B.** dilation of $\triangle DEF$ by a scale factor of 2 followed by a reflection across the y-axis
- **C.** dilation of $\triangle DEF$ by a scale factor of 3 followed by a reflection across the x-axis
- **D.** dilation of $\triangle DEF$ by a scale factor of 3 followed by a reflection across the y-axis

52. A survey of students in a class explored the relationship between gender and which of three summer Olympics sports students most enjoyed watching.

	Gymnastics	Track and Field	Soccer	Total
Boys	2	6	7	15
Girls	9	1	5	15
Total	11	7	12	30

If a girl is chosen from the class at random, what is the probability that her favorite sport to watch is gymnastics?

_____

53. Estimate the value of $2\sqrt{2}$ to the nearest tenth.

_____

54. Solve for z: $\frac{3}{4}z - 1 = \frac{1}{4}(z + 8)$.

_____

55. The cylindrical can shown below has a height of 10 inches and a volume of 250π cubic inches. What is its radius, r?

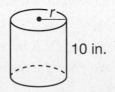

10 in.

_____

56. What is the length, in units, of $\overline{MN}$ on the coordinate grid below?

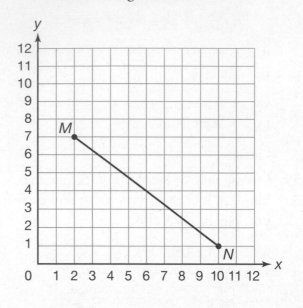

_____

57. The diameter of a presidential $1 coin is 0.002 meter. A bacterium has a diameter of 5×10^{-7} meter. About how many bacteria that size would fit across the diameter of the coin?

_____

STOP

Session 2

58. Aiden wants to rent rollerblades when he goes to the park. He also needs to rent a helmet. Rates for two rental shops are shown below.

Skate Heaven
$3 for the helmet plus $2 per hour for rollerblade rental

Rent-a-Rama
$1 for the helmet plus $3 per hour for rollerblade rental

A. Let y represent the total cost, in dollars, of renting rollerblades and a helmet.

Let x represent the number of hours for the rental.

Write a system of equations to represent this problem situation. Then graph the system on the coordinate grid below.

_____

_____

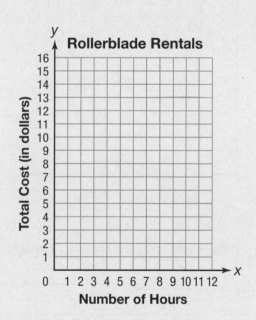

Rollerblade Rentals

Total Cost (in dollars)

Number of Hours

B. For how many hours would Aiden need to rent rollerblades in order for the total cost to be the same at both shops? What would that total cost be? Use the solution for the system of linear equations to explain your answers.

_____

_____

_____

59. The table below shows the number of years of experience that 8 employees at a supermarket have and their hourly wages.

Experience and Hourly Wage

Years of Experience	1	2	2	3	4	5	5	9	9	10
Hourly Wage (in dollars)	$8	$8	$10	$11	$10	$11	$12	$14	$15	$15

A. Create a scatter plot of these data on the grid below. Choose a scale that allows you to plot all the data. Be sure to title your scatter plot and label each axis. Then draw a line of best fit for the data.

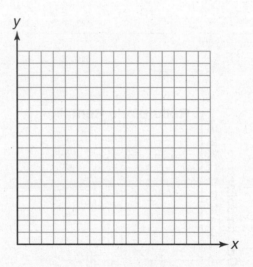

B. Describe the association shown by the scatter plot. Explain how you know.

_____

_____

_____

60. Look at △ABC and △HJK below.

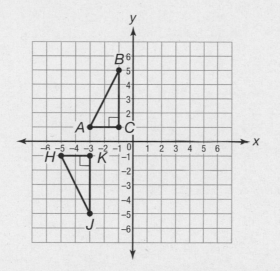

A. Describe how you could use a sequence of two rigid transformations (reflections, rotations, and/or translations) to move △ABC so that it completely covers △HJK.

_____

_____

_____

B. Do the transformations in Part A prove that the two triangles are congruent, similar, or both? Explain how you know and list all the pairs of corresponding sides and corresponding angles for these two triangles.

_____

_____

_____

STOP

Math Tools: Grid Paper

Math Tools: Coordinate Grid

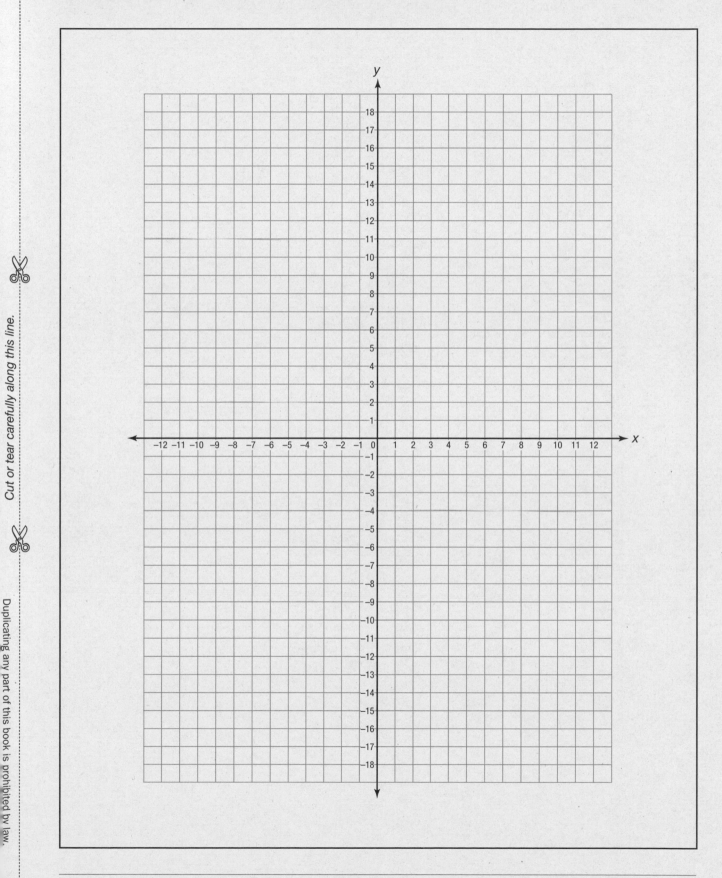

Math Tools: Grid Paper

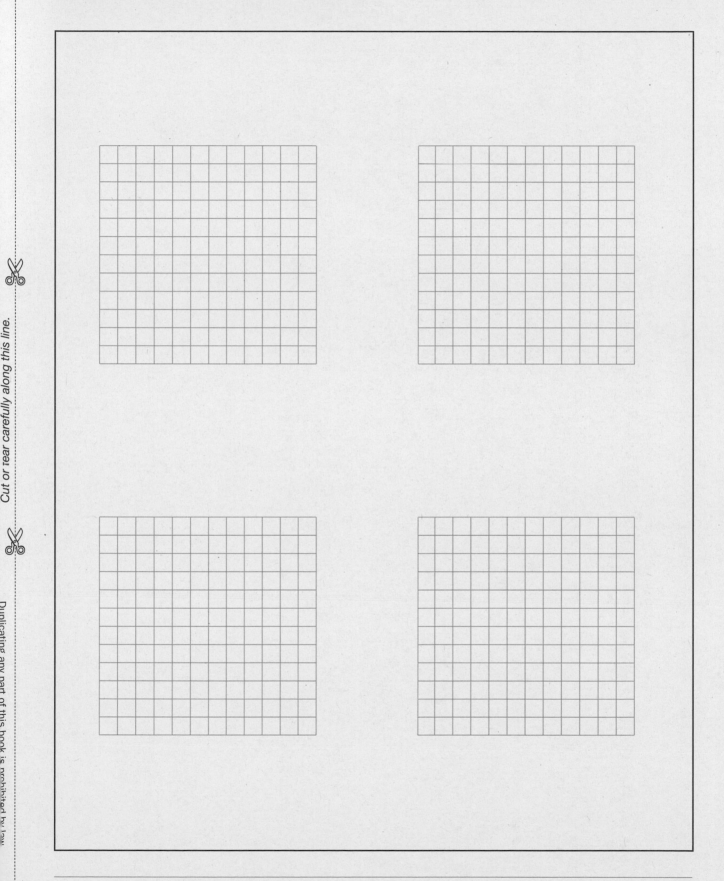

Math Tools: Coordinate Grids

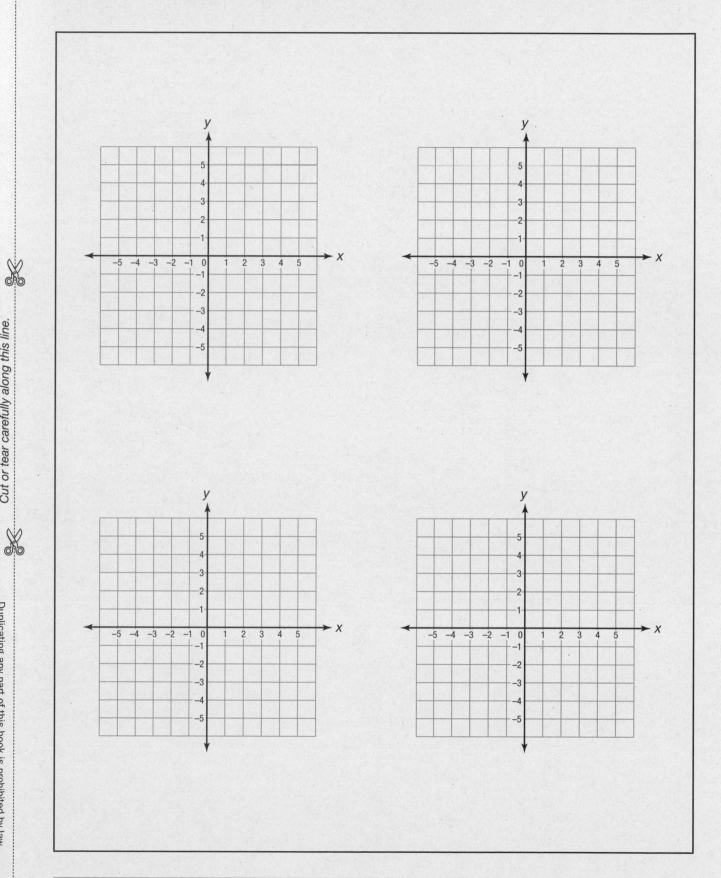